The Fifth Level of Evolution

Manuel Alfonseca

© **Manuel Alfonseca, 2020**

ISBN-13: 979-8603138909

All rights are reserved. Apart from any fair dealing for the purpose of private study, research, criticism or review, as permitted under the Copyright Act, no part of this publication may be reproduced, stored in a retrieval system, or transmitted, in any form or by any means, electronic, electrical, chemical, optical, printing on paper, photocopying, recording or otherwise, without the previous permission of the copyright owner.

TABLE OF CONTENTS

1. Evolution of the universe before the apparition of life — 5
2. The first level — 21
3. The second level — 39
4. The third level — 53
5. The fourth level — 57
6. What is man? — 75
7. Towards the fifth level — 97
8. The fifth level in literature — 115
9. The Omega point — 139
10. Internet as a nervous system — 157
11. Must we renounce reproduction? — 179
12. Can we control our evolution? — 203
13. Should we control our evolution? — 225
14. How will the fifth level be? — 245
15. Does the fifth level exist? — 275

Bibliography — 293

1. Evolution of the universe before the apparition of life

At the beginning of every great cosmic cycle, Vishnu is in eternal rest. From his belly button, a lotus flower is born, and from the flower comes Brahma, the creator. Brahma extracts from the waters the cosmic egg, the universe, which is divided into two halves, each of which comprises seven strata. The first stratum in the upper half is the Earth where we live, in whose center rises Mount Meru, the universal mountain, through which passes the axis of rotation of the cosmic egg. The central circular continent of the Rose spreads around the mountain, surrounded successively by a sea of salt water, an annular continent, a sea of sugar, another annular continent, a sea of milk, and so on, up to seven seas and seven continents.

We call *cosmology* the science that studies the properties, origin and evolution of the universe. Since antiquity, as well as in the most primitive cultures, man has been worried by these issues and has formulated answers that today may seem naive, where the facts within the reach of our knowledge were mixed with mythical-religious elements, in order to produce a coherent whole. The preceding paragraph describes one of those cosmologies, which appeared in India half a millennium before Christ.

As man's knowledge of the world increased, the proportion of religious elements in cosmological constructions decreased, to the point that, during the twentieth century, cosmology has become a branch of the physical sciences.

Before describing what modern science says about the origin and evolution of the universe, let us consider the historical process though which it was concluded.

We will begin in ancient Greece, in *the myth of Perseus*, who at the request of King Polidectes undertook the risky task of conquering the head of the gorgon Medusa, who turned into stone anyone who looked at her. After achieving his goal with the help of Hermes and Athena, Perseus undertook long journeys. In one of them he arrived in Ethiopia, where reigned king Cepheus, husband of Cassiopeia, whose daughter, Andromeda, was incomparably beautiful. The queen made the mistake of boasting of her daughter's beauty, declaring her superior to the Nereids, divinities of the sea. Poseidon, god of the waters, furious at Cassiopeia's daring, sent a sea monster to ravage the coasts of the country. An oracle told Cepheus that the scourge would cease if he gave his daughter to the monster. The king ordered Andromeda to be chained to a rock near the beach, and leave her at the mercy of the beast. But Perseus, riding the winged horse Pegasus, came to her aid, killed the monster, saved the girl and married her.

The Greeks used their myths to name the constellations they saw in the night sky, where their fertile imagination perceived complicated drawings that represented some of the characters in their mythology. In particular, in the northern part of the celestial sphere, five nearby constellations received the names of the myth we have just summarized: Cepheus, Cassiopeia, Perseus, Pegasus and Andromeda.

The Fifth Level of Evolution

* * *

After the collapse of the Roman Empire of the West, while Western civilization was sunk in the semi-darkness of the early Middle Ages, the young Islamic civilization assimilated and increased the astronomical knowledge of the Greeks. In a catalog of stellar objects (*The Book of Fixed Stars*), compiled by the Persian astronomer Abd-al-Rahman al-Sufi (903-986), we can find the first mention of a certain mysterious object, similar to a bright cloud and located in the Andromeda constellation.

In 1611, the German astronomer Simon Marius (1573-1624) was the first to observe the Andromeda nebula through a telescope. Much later, in 1845, the British scientist William Parsons (Lord Rosse, 1800-1867) discovered that some objects similar to this nebula have a curious spiral structure. The first photographs of the Andromeda Nebula, obtained in 1885 by Isaac Roberts, proved that it was a member of the class of spiral nebulae.

The German philosopher Immanuel Kant (1724-1804) and the German-British astronomer William Herschel (1738-1822) had formulated the theory that the nebulae could be clusters of millions of stars, similar to the region of the universe where we are. They proposed to give them the name of island universes. However, one century later most astronomers felt skeptical.

Before going further, let's get back a few centuries and look at the advances in a different science, the part of physics that studies the properties of light, which is called *optics*[1].

---

[1] From the Greek *optos*, visible.

Since ancient times it was well known that when white light passes through a drop of water or a properly shaped glass, iridescent colors appear. But it was the English physicist Isaac Newton (1642-1727) who proved that white light is made of the mixture of a continuous succession of colors, which is called a *spectrum*[2].

In 1802, the English chemist William Hyde Wollaston (1766-1828) discovered that the spectrum of sunlight is not continuous, for it shows many dark, irregularly distributed stripes. In 1859, the German physicist Gustav Robert Kirchhoff (1824-1887) showed that the gases of various chemical elements absorb certain wavelengths (colors) of light, which means that the stripes in sunlight have been made by absorption of the corresponding wavelengths by the elements in the sun's chromosphere. In addition to explaining the mystery of the dark bands, Kirchhoff's experiment made it possible to discover the composition of the chromosphere. Applying the same method to other stars, their composition can also be deduced. So it was found that hydrogen and helium are the most abundant elements in the universe. In fact, helium was unknown on Earth until its stripes were found in the solar spectrum. That's where its name comes from, for the Greeks called the sun *Helios*.

We now come to the third link we need to continue with the history of the discovery of galaxies. We must now to another branch of physics: *acoustics*, which studies sounds.

A well-known effect takes place when we are overtaken by a vehicle that generates a continuous sound, such as a car, when the driver is blowing the horn. The sound seems to undergo a sharp

---

[2] From the Latin *spectrum*, image.

change in tone, suddenly becoming deeper when the vehicle passes us. In 1842, the Austrian mathematician and physicist Christian Doppler (1803-1853) studied these phenomena and formulated a law that makes it possible to calculate the frequency shift produced in sound waves, as a function of the speed of the object generating them. The higher the speed, the greater the pitch shift.

An important consequence of the mathematical formulation of the *Doppler Effect* is that, if one knows the frequency change in the waves, one can calculate the speed of the body producing them. It should be added that the same effect applies to all kinds of wave phenomena: not just sound, but also light.

The Doppler Effect has practical applications: when a star moves away from us, its light decreases in apparent frequency (increases in wavelength). If we get the spectrum of that light, we'll see that the dark lines of the various elements are displaced towards the region of greater wavelength (the red area in visible light). That's why we say that the spectrum has suffered a redshift. The opposite takes place when the star is coming near: we speak of a blue shift. The radial velocity of the star can be deduced from the amplitude of the shift.

* * *

We can now go back to the Andromeda nebula. In 1913, the American astronomer Vesto Melvin Slipher (1875-1969) obtained its spectrum and discovered a blue shift that indicated that the nebula is moving towards us with a speed of about 300 kilometers per second, much greater than was expected. Slipher then studied the light of other spiral nebulae and made the unexpected discovery that most of them, unlike Andromeda, show redshifts, that is, they are moving away very quickly from the solar system. He found speeds above 1000 kilometers per second.

In 1919, the American Edwin Powell Hubble (1889-1953) used the Mount Wilson telescope to photograph several spiral nebulae, including Andromeda, and proved that they were really huge groups of stars. The theory of the island universes was thus confirmed. From then on they were no longer called nebulae, but galaxies, in honor of our Milky Way[3], which also belongs to the class of spiral nebulae.

Hubble's interest in the Andromeda galaxy did not stop there. He wanted to calculate its distance from us. To do this, he relied on a discovery made in the constellation of Cepheus, neighbor of Andromeda's and father of the princess, according to the Greek myth.

In various places in the sky, but especially in the constellation of Cepheus, where they were discovered, there are stars whose light varies regularly in intensity, therefore they are called *Cepheid variables*. In 1908, Henrietta Swan Leavitt (1868-1921) discovered that the period of variation of these stars is linked to their luminosity. The higher the latter, the longer the period.

How could this help us measure the distance separating us from the Andromeda galaxy? Some Cepheid stars are quite bright, and Hubble located about forty in the photographs of that galaxy. After measuring their periods of variation, he applied the function found by Leavitt to get their real luminosity. From its comparison with their apparent luminosity, the distance can be deduced[4].

---

[3] *Galactos* in Greek means milk.
[4] The apparent luminosity of an objects gets lower in inverse relation to

Modern calculations, more accurate than those made by Hubble, allow us to assert that the Andromeda galaxy is more than two million light-years from Earth, i.e. about 18 quintillion kilometers, and the light we now receive from that galaxy was generated more than two million years ago, more or less at the time of the apparition of man on Earth.

Hubble's experiments did not stop there. Continuing Slipher's work and with the help of Milton Lasalle Humason, he measured the speeds of many other galaxies and calculated their distance. In 1929 he published his famous law, which in fact had been computed by Georges Lemaître two years before: *The farther a galaxy is, the faster it moves away from us.* In other words: the universe is expanding. Galaxies are like the points marked on the surface of an inflating balloon.

From then until now, the Hubble-Lemaître's law has been confirmed. Today we have found very distant galaxies and stellar objects, billions of light-years away, with redshifts so large that some of these objects seem to move away from us (according to the Doppler Effect) with speeds close to the speed of light.

However, the idea that the universe could be expanding did not take scientists by surprise. In 1917, Albert Einstein (1879-1955) published his general theory of relativity, which leads to an equation that predicts the variation of the volume of the universe as a function of time. Depending on the values of its parameters, various solutions can be obtained. Some of these describe an expanding universe, but until Lemaître and Hubble formulated their law, no one had taken them seriously.

---

the square of the distance.

* * *

The fact that the cosmos is expanding like the surface of an inflating balloon has a very important consequence: at some point in the past it must have been very small. In other words, the expansion had to begin at a certain point in time. Our universe, as we know it today, had a beginning.

Some scientists felt uncomfortable with this situation. If the cosmos had a beginning, perhaps we should accept that science has limits: the origin of time and of the universe would escape from human knowledge. On the other hand, they were afraid that science was introducing the need for a transcendent action: a creator. Aesthetic reasons were also adduced to oppose the theory of the primitive expansion (the *Big Bang theory*), such as the fact that, if most galaxies really move away from us, there will come a time when we can no longer see any, and then we'll be alone, lost in an infinitely empty universe.

It may seem curious that such arguments can be used to justify a cosmological theory, but we must bear in mind that scientists are human beings, subject to partially irrational prejudices and impulses, like everyone.

In 1948, British astronomers Hermann Bondi and Thomas Gold proposed the theory of the *steady-state universe*, later supported by the British astronomers Fred Hoyle and Raymond A. Lyttleton[5]. According to this theory, the average density of the universe remains constant despite its expansion. To achieve this, it was necessary to renounce the principle of energy conservation, the most sacred in physics. The steady-state theory asserts that matter

---

[5] Raymond Lyttleton, *The Modern Universe*, 1956. Arrow Books, 1960.

is continually created spontaneously, in exact proportion to compensate for the distancing of galaxies. Over billions of years, the created matter will be grouped into new galaxies that will replace those that have moved away, so there will be no danger of ever feeling alone.

For seventeen years, the steady state theory achieved great favor among physicists and astronomers. But suddenly, unexpectedly, a phenomenon was discovered that could only be explained by the assumption of the initial expansion of a super-compact universe. The steady state theory was abandoned even by its creators. After all, scientists are reasonable.

That clarifying phenomenon is the microwave cosmic background radiation, discovered in 1965 by Robert W. Wilson and Arno A. Penzias, who in 1978 received the Nobel Prize. The finding came when the radio telescope with which they were working detected a strange background noise in the region of the radio spectrum corresponding to microwaves. After eliminating all the possible causes of terrestrial origin, it was concluded that it must come from an extraterrestrial source.

This radiation has very peculiar characteristics: it comes equally from all directions of space; it varies with frequency almost identically to the emission of a black body (a perfect radiator) at a very low temperature: about three degrees above absolute zero. Its existence had been predicted sixteen years earlier by Ralph A. Alpher and Robert C. Herman, starting from the hypothesis that the theory of the *Big Bang* is correct. The triumph of this theory was, therefore, complete.

* * *

Today, several decades after the discovery by Wilson and Penzias, the theory has been refined to describe in some detail the events that took place at the origin of time. This is what probably happened:

At the beginning, in the initial moment of the great expansion, the universe was incredibly compressed at a very high temperature and pressure, but we don't know anything about what happened in the first 10^{-43} seconds (Planck's time), because the two basic theories of current physics (quantum mechanics and relativity) would have to be applied simultaneously before that time, and it turns out that they are contradictory, we need a new unified theory.

This does not mean that there are no theories trying to explain what could have happened before the beginning of the universe (if that phrase means something, for time, like space, is a property of the universe and probably does not exist outside the universe). There is talk of multiple universes, of the M theory, and the worst is, that their defenders present these things as if they were science, when in fact they are mathematical and metaphysical elucidations without experimental basis. It is obvious that we cannot perform experiments outside the universe, and that is what should be done to prove those theories. Actually, the worst is not that we cannot verify them, but that they cannot be proven to be incorrect. That is precisely why they are not scientific[6].

The reason why many cosmologists let their imagination run away, and invent theories about what could have happened before the beginning of the universe is, at bottom, religious (or perhaps we should say antireligious). Since the mid-twentieth century, atheism

---

[6] Karl R. Popper, *The logic of scientific discovery*, 1934.

is on the defensive against modern cosmological theories, and holds on to anything, to deny God's existence. For some atheist cosmologist, the way of escape was for some time the steady state theory, but as this theory was abandoned, they have found increasingly strange and less scientific alternatives, justified because the mathematics are correct, apparently forgetting that mathematics can describe completely different universes, while we are certain of the existence of just one: ours.

At Planck's time, the average density of the universe was 10^{94} times greater than that of water. The entire cosmos was concentrated in a volume similar to that of an atomic nucleus. Atoms did not exist. According to a version of the *Big Bang theory* called the *inflationary theory*, due to Alan Guth, the expansion would have been much faster at first, but this phase would have lasted a very small fraction of a second.

Gradually, as the expansion progressed, elementary particles began to emerge, in other words, matter. First there was a quark soup, and when the quarks could join together, protons, neutrons, electrons and neutrinos, along with their antiparticles and more exotic particles that appeared and disappeared continuously. The density was so high that a particle could not travel a long distance without encountering its corresponding antiparticle. When this happened, both were annihilated, transforming into photons, i.e. energy.

But the expansion continued. A thousandth of a second after the beginning of time, the volume had increased to such an extent that protons and neutrons could no longer originate spontaneously. From that point, the components of matter have been more or less fixed.

As the universe expanded, the pressure, temperature and density decreased. One second after the *Big Bang*, the temperature had dropped to ten billion degrees, while the average density was approximately equal to that of water. A minute later, the temperature had dropped to a few hundred million degrees. At that time, the conditions favored the fusion of free protons and neutrons to form more complex nuclei: deuterium[7] and helium[8]. In a few moments, about twenty percent of matter was transformed into helium, which for this reason is now the second most abundant element in the universe[9]. The production of helium stopped when the temperature dropped below one million degrees, about three minutes after the initial explosion.

Let us now take a leap forward: three hundred and eighty thousand years after the origin of the cosmos. The temperature has dropped to 3000 Kelvin. At this time, the expanding universe left us the signature or seal of its origin in the form of the cosmic background radiation. Let's see how it happened.

We know that an atom consists of a nucleus, made of protons and neutrons, and a number of electrons, equal to that of protons in the nucleus. However, at temperatures of several thousand degrees, electrons are torn from their orbits by continuous collisions between atoms, which at those temperatures move at high speed. Matter is not then made of neutral atoms, but of a mixture of positively charged nuclei and free electrons. This is called the *plasma state*.

---

[7] A heavy form of hydrogen made up by a proton and a neutron.
[8] Its nucleus is made of two protons and two neutrons.
[9] The first is hydrogen, whose nucleus is made of a single proton.

Light cannot pass through a mass of plasma, for photons are captured by the nuclei and the free electrons. Therefore, plasma is opaque, while ordinary gases are usually transparent, as they are formed by neutral atoms, which don't react easily with light.

When the universe cooled below 3000 degrees, the atomic nuclei were able to capture electrons without their being removed almost immediately. Many neutral atoms suddenly formed, and most of matter passed from the plasma state to the gas state. The cosmos, which up to then had been opaque, became transparent. This happened so suddenly, that it must have had the effect of a huge blinding flash of light.

The expansion has continued for over thirteen billion years, and some of the light rays that were then originated are just now reaching us. From where they come?

Obviously, from points now located over thirteen billion light years from Earth, for that's the distance traveled by light in that time. But, according to the law of Hubble-Lemaître, the redshift of light coming from distant parts of the expanding universe is proportional to their distance from us. If we apply this law to the light coming from the initial flash of light that is now reaching our galaxy, its frequency will have decreased so much, that these electromagnetic waves can no longer be considered visible light, but microwaves. It can also be proved that this light converted into microwaves should be identical to the radiation of a black body at a temperature of a few degrees above absolute zero.

All these characteristics belong to the cosmic background radiation. Therefore, this high frequency radio noise that invades everything must be precisely the residue of the initial flash of light we were looking for. The photons arriving to us now started their trip less than four hundred thousand years after the origin of the universe. They are, therefore, very old. In fact, they are the oldest trace we can directly see of the origin of the universe, for the cosmos was opaque before that moment, and no electromagnetic signal coming from it can reach us now.

The reader will remember that the cosmic background radiation and its representation, detected by radio telescopes located on satellites, often appear in newspapers. Although, in general, as I use to say to my students, scientific news in the press are unreliable, those photographs can be considered reliable. They indicate that the cosmic background radiation exhibits minute temperature changes depending on the direction, which shows that the early universe was not perfectly uniform. Small heterogeneities must have appeared quite early: a greater density of matter here, a thinning there... In the areas of greater density the gravitational field acted more intensely (the more mass, the more attraction; the more attraction, the more mass). Heterogeneities were amplified. As a result, the universe acquired granular structure and matter condensed in lumps, separated by vast empty spaces. Those lumps are the primitive galaxies, whose formation took place around five hundred million years after the origin of the universe.

Although the distance between the galaxies is very large, the effect of gravity on the nearest ones counteracts the expansion of the universe, so these gigantic celestial objects tend to make clusters, some of them enormous. The galaxies in a cluster don't move away from each other and may get quite near in the course of their displacements, as will happen with Andromeda and the Milky

Way. In some cases two galaxies have collided, with incredibly catastrophic effects.

Once the galaxies were formed, their matter was not distributed regularly within their volume. Soon there were density inequalities, clouds of gas and, finally, the first stars were born. Phenomena like these occur now before our eyes in the Milky Way galaxy.

2. The first level

About 4600 million years ago, a cloud of gas and dust started contracting. In the center of one of its condensation nuclei, a new star appeared. This phenomenon, which has been repeated countless times in the history of our galaxy, was especially significant for us, because that star was the Sun.

At various distances from the central star, near the equatorial plane of rotation of the nebula, there were secondary condensations that gave rise to the planets of the Solar System. In the regions closest to the Sun, the temperature was high; thus only the less volatile substances (iron, silicates) became viable conglomerates. Today they are the planets Mercury, Venus, Earth and Mars, plus the Moon and many asteroids.

Three of these stars (Venus, Earth and Mars) grew enough to retain, by gravitational attraction, an atmosphere rich in volatile substances, such as water, nitrogen, methane and carbon dioxide. At the prevailing temperature in the interior regions of the Solar System, they could not retain the lightest gases, hydrogen and helium, except in combination with heavier elements. Helium, the most inert of the chemical elements, does not combine with any other and is practically absent from Earth. Hydrogen, on the other hand, is very reactive and easily bonds with oxygen, nitrogen and carbon, relatively abundant in the original nebula. The compounds thus formed are water (H_2O), methane (CH_4) and ammonia (NH_3). It is assumed, therefore, that the primitive terrestrial atmosphere could be made of a mixture of these gases, together with carbon dioxide (CO_2) and molecular nitrogen (N_2).

The planets of the outer Solar System, Jupiter, Saturn, Uranus and Neptune, were formed in colder regions, further from the condensation core. At temperatures not much higher than absolute zero, the molecules of light gases scarcely move and can be retained by the gravitational attraction of Earth-sized bodies. It is believed that the giant planets have in their center a solid core of this type, around which hydrogen, helium and the remaining gases accumulated. In this way, these planets retain a composition similar to that of the primitive cloud, while in the inner Solar System there was a large concentration of heavy elements and a dearth of lighter more abundant atoms.

The primitive Earth's interior trapped radioactive elements, such as uranium and thorium, which came from the explosion of nearby supernovae. Their proportion was then greater than it is today, for they have gradually disintegrated over billions of years and cannot be replenished. When they disintegrate, they release energy, which heats the Earth's interior. There was so much radioactivity during the first millions of years of the existence of our planet, that it became very hot. All available water was then in a gaseous state, making an atmosphere much denser than today, while the surface of the Earth melted or, at least, was viscous.

Another phenomenon that contributed to the warming of the early Earth was an intense bombardment of meteorites. When the Solar System was young, the number of small bodies it contained was much greater than it is today. The probability that any of them collided with one of the big bodies was therefore quite high. We can see on the Moon the scars of meteor impacts, whose approximate age can be calculated by statistical methods. Studies show that, over 4 billion years ago, the lunar surface was hit at a rate thousands of times greater than today. Over time, the number of meteorites declined. About 3 billion years ago, the rate of formation of lunar craters dropped to values close to current ones.

The data obtained on the Moon have been confirmed through the modern exploration of more distant bodies whose surface is also marked with craters: Mercury, Mars and a few satellites. It is clear that the Earth was not exempt of the bombardment, but our craters have mostly disappeared, obliterated by erosion and the movements of the continents. Nevertheless, traces of violent impacts have been discovered in old terrestrial regions that have been relatively stable for billions of years. One of these regions makes most of Canada.

After a few hundred million years, radioactive elements of shorter life were running out, impacts of meteorites became less frequent, and the Earth cooled. 4100 or 4200 million years ago, the first solid rocks of the crust appeared, such as those discovered in Australia. Other less ancient, found in Greenland, with an age of 3950 million years, are associated with sediments of the same age, which shows that at that time water already existed on Earth in a liquid state.

* * *

It is believed that life appeared on Earth quite early, just a few million years after the crust was solidified and the oceans appeared. The crucial experiment that opened the way to current ideas regarding the origin of life took place in 1952 and was carried out by the American scientist Stanley Lloyd Miller, a collaborator of Harold C. Urey, who filled a container with a mixture similar to what was then supposed to have been the primitive atmosphere (methane, hydrogen, ammonia and water), closed it tightly and subjected it to electric sparks during a week. The analysis of the resulting mixture discovered the presence of simple organic substances that pointed the way to the appearance of life. The most notorious compounds belonged to the group of amino acids, the fundamental blocks of proteins.

If the primitive Earth had had an atmosphere like the mixture used by Miller, similar chemical reactions would have taken place. The energy source could have been provided by atmospheric electric discharges in storms; thermal energy would have been released by volcanic eruptions; by hot springs coming up from the Earth's mantle in the areas of separation between the plates of the Earth's crust, in the deep sea; by decomposition of radioactive substances; finally, by the sun's ultraviolet light. Of these five sources, the first two appear irregularly and have variable intensity: we don't have storms every day, and volcanic eruptions are relatively rare, although perhaps they were more frequent four billion years ago.

Hydrothermal vents, radioactivity and ultraviolet light are, on the contrary, practically constant sources of energy. The last two were then much more abundant than they are today. The primitive atmosphere of the Earth was transparent to short wavelength solar light, which is now absorbed by the ozonosphere, the ozone layer that envelops the Earth at a height of 30 to 40 kilometers, which then did not exist. Miller's experiment has been repeated by other researchers, using ultraviolet rays instead of electric sparks, with similar results. It is evident, therefore, that enormous quantities of simple organic substances could have appeared spontaneously on the primitive Earth.

But the presence of these chemical compounds was not enough. It was necessary that they were protected after their formation against destruction by the same energy sources that made their origin possible. Organic substances are unstable. High temperatures and ultraviolet rays break them down into their constituent elements. In the absence of protection, a balance would have been reached where organic matter would be destroyed as quickly as it was created. Something like this has possibly happened on Mars, on whose surface the activity of ultraviolet rays is so great, that the analyzes carried out by the Viking space capsules 1 and 2 did not discover clear traces of organic matter, not even what was expected as a result of meteorite impacts, for substances of this type have been produced by spontaneous reactions in meteorites.

Why didn't the same thing happen on Earth as on Mars? Because the Earth, unlike our neighbor planet, has oceans. Part of its surface is covered by water in a liquid state. A large part of the spontaneously produced organic substances fell into the sea and sank down. The three-dimensional mass of the ocean provided them with a medium where they could move freely and react easily, while water molecules acted as a screen that protected them from ultraviolet light.

It is estimated that organic matter accumulated in the ocean to such a point, that its concentration reached one percent. As the weight of ocean water exceeds one quintillion tons, if the previous figure is correct, the amount of organic matter was enormous. This is the reason why the primitive ocean is called the *primordial soup*.

In experiments carried out after 1952 to simulate the processes that took place on Earth during the beginning of its history, the composition of the mixture has been changing: substances obtained as the result of an experiment were introduced from the start in the next step, as it was reasonable to assume their existence on the primitive Earth. This is how the Spanish chemist Juan Oró (1923-2004) started in 1961 with mixtures containing hydrocyanic acid, one of the compounds obtained in 1952. After subjecting the mixture to energy, he discovered the presence of adenine among the resulting products. A year later he added formic aldehyde to the initial composition and found two sugars: ribose and deoxyribose.

* * *

If the amino acids obtained in 1952 by Miller opened the way to proteins, the three substances obtained by Oró gave the first link in the chain leading to other kind of huge and complex molecules, *nucleic acids*, which are a fundamental part of all living beings and have such special properties that today it seems possible to consider them *alive*.

The molecules of nucleic acids have a linear structure, formed by the assembly of short units, called nucleotides. The German biochemist Albrecht Kossel (1853-1927) received in 1910 the Nobel Prize in Physiology and Medicine for the discovery of the chemical structure of these elementary blocks. Each one of them is made, in turn, of the union of three simpler substances: phosphoric acid (PO_4H_3); a sugar; and a nitrogenous base, derivative of purine or pyrimidine. Depending on the base and the sugar, there are eight different types of nucleotides.

The sugars that make a part of the nucleotide have five carbon atoms and are therefore called pentose[10]. Its structure is similar to that of ordinary honey sugar, glucose, which has six carbon atoms and therefore belongs to the group of hexoses[11]. Two different forms of pentose may appear in the nucleotides of the nucleic acids. The first, ribose, was synthesized in 1901 by the German biochemist Emil Fischer (1852-1919), who the following year received the Nobel Prize in Chemistry. The synthesis of ribose took place before its presence in nucleic acids was discovered, which did not happen until 1908. Its reduced formula is $C_5H_{10}O_5$. The second pentose is deoxyribose, identical to ribose except for the lack of an oxygen atom (hence its name). Its reduced formula is $C_5H_{10}O_4$. As for nitrogenous bases, there are five different types: adenine (one of the substances obtained by Oró), guanine, cytosine, thymine and uracil.

We have seen that a nucleotide consists of a molecule of phosphoric acid combined with a pentose and a nitrogenous base. Since there are two different kinds of pentose, five bases and a single type of phosphoric acid, in theory ten different combinations could be formed. However, two of them do not occur in the natural state: ribose with thymine and deoxyribose with uracil.

To form a nucleotide, pentose combines on the one hand with phosphoric acid and on the other with the nitrogenous base. The result is the following chain: Phosphoric-Pentose-Base. Several

---

[10] The ending -*ose* is used in Chemistry to name carbohydrates or sugars; the Greek word *penta* means five.
[11] From the Greek *hexa*, six.

nucleotides can bind to each other through bonds that bind the phosphoric acid of one with the pentose of the other. This process can be repeated many times, which results in very long chains: nucleic acids. The first artificial performance of such a reaction won the 1959 Nobel Prize in Physiology and Medicine for Severo Ochoa.

Nucleotides do not bind indiscriminately to form nucleic acids. There is an important restriction: either all of them contain ribose, in which case the resulting complex is called ribonucleic acid (RNA in short), or they all contain deoxyribose, giving rise to deoxyribonucleic acids (DNA). There are, however, no restrictions on the bases hanging from the links in the chain. Any base, of the four possible in each case (adenine, guanine, cytosine and uracil for RNA; adenine, guanine, cytosine and thymine for DNA) can appear in any place, which means that there is no single nucleic acid of each type, but a huge number of possible combinations.

How many? Let's make some calculations. The DNA molecule of a virus can contain several thousand nucleotides. The figure increases to several million in bacteria, and to billions in mammalian cells. In 1976, the first complete genome was obtained: the sequence of bases in the DNA of a small virus, ΦX174, whose molecule contains 5375 nucleotides. Each of the 5375 nitrogenous bases could have belonged to one of the four species indicated above: adenine, guanine, cytosine and thymine, as all are interchangeable. It turns out that the DNA molecule of the ΦX174 virus is just one among all possible combinations that could be formed with those four bases in chains of 5375 elements. That number of combinations is equal to $4^{5375} = 1.2 \times 10^{3236}$, a number made of the digits one and two followed by 3235 zeros.

The Fifth Level of Evolution

Fifty-four lines of an ordinary book would be necessary to write it, just over a page and a half.

The reader who has experience with mathematics will not find it difficult to understand the enormous magnitude of these numbers. For those who don't have it, table 2.1 can give an idea of how quickly exponential numbers grow. Recall, in addition, that ΦX174 is one of the smallest viruses. The degree of variety contained potentially in the structure of natural beings is impressive.

Number	Name	Examples of approximate magnitudes
10^6	Million	Number of inhabitants in a great city
		Number of copies of a large-edition newspaper
		Yearly world production of gold in kg
		Number of different ways to order 10 persons in a row
10^{12}	Trillion	Yearly world production of carbon (kg)
		Yearly world production of natural gas (m^3)
		Number of different ways to order 15 persons in a row
10^{24}	Septillion	Age of the universe in millionths of a second
		Number of molecules in 33 liters of a gas
		Total weight of all the ocean water in grams
		Number of different ways to order 24 persons in a row
10^{48}		Number of different ways to order a deck of cards
		Number of elementary particles in the oceans
		Number of different ways to order 40 persons in a row
10^{80}		Number of elementary particles in the visible universe.
10^{158}		Number of different ways to order 100 persons in a row
10^{375}		Number of different ways to order 200 persons in a row
10^{2567}		Number of different ways to order 1000 persons in a row

Table 2.1. Examples of exponential growth

We have seen that DNA nucleotides form long chains in alternating links of phosphoric acid and pentose, from which purine and pyrimidine bases hang. In practice, the resulting chain does not usually follow a straight line.

In addition, nitrogenous bases can also be linked together by means of a less energetic chemical bond, called a hydrogen bridge, for it is carried out by means of an atom of this element. Adenine can thus form a bond with uracil and thymine, while guanine can bind with cytosine. This means that a chain of a nucleic acid can form a relatively stable structure by joining with another chain whose bases are complementary (meaning that they can form hydrogen bonds). For example, the following two chains are complementary:

```
G A T T A C A
C T A A T G T
```

where A = Adenine, G = Guanine, C = Cytosine, T = Thymine. If we have the complementary of an entire chain, both will make a single body, winding into a double helix whose shape resembles that of a spiral staircase with steps. The structure of DNA was discovered by the American biologist James Dewey Watson and by the British Francis H. Compton Crick and Rosalind Franklin. The first two received the Nobel Prize in Physiology and Medicine in 1962.

If the two chains of DNA separate, because the hydrogen bonds between their bases have been broken, each of them is capable of generating its complementary chain from elementary blocks (nucleotides), through mechanisms whose description we won't describe here. In this way, where we had a single macromolecule of nucleic acid, we now find two exact copies. In other words, the molecules of these organic compounds are able to reproduce.

* * *

We have thus an interesting question: are nucleic acids alive? Which immediately leads to another previous question: What is a living being? What's the definition of life? Before reading the following paragraphs, I suggest that the reader will try to answer this question by himself. If you think about it, you'll see that it's more complicated than it seems at first glance.

According to the traditional philosophical interpretation of the medieval Greco-Roman and Western civilizations, a living being is an organized entity endowed with continuous and immanent movement, in other words, a complex being formed of simpler parts assembled together, capable of moving by itself.

This definition of life is not satisfactory. There are machines created by human technology that should be considered organized, complex and composed of simple parts. Some are able to move by themselves. However, no one claims that a space capsule or a drone is alive.

Secondly, we have no doubt that each of the cells that make up the human body is alive, despite the fact that many cannot move during most of their lives. Bone cells, for example. The above definition of life must be incorrect, as it doesn't apply to some living beings, and it does apply to others that are not living. We must find a better definition.

A somewhat modern definition, accepted by biologists of the nineteenth century, defines a living being as an entity that is born, grows, nourishes, reproduces and, sometimes, dies. The restriction applied to death is due to the existence of many unicellular beings (and a few multicellular) capable of dividing into two different and independent organisms, which continue to live and reproduce indefinitely, never suffering a true biological death, unless they have an accident or become the prey of a predator.

Those biologists defined life based on the activities of living beings that differentiate them from inert ones. It is, therefore, a functional, non-structural definition, and the distinctive characters chosen are normally grouped into three functional categories: nutrition, relationship, and reproduction. The first allows the living being to exchange matter and energy with the surrounding environment; the second enables it to react to the stimuli received from that environment; the third results in the appearance of new living beings that replace or are added to existing ones.

The enormous development achieved by Western technology since the industrial revolution has led to the emergence of increasingly complex machines, which often work analogously to some living beings. There are electronic computers capable of extracting energy from the surrounding environment, which have sensors that allow them to measure certain physical-chemical variables and produce responses that depend on the values they have measured. It can be said that these machines are capable of performing nutrition and relationship functions similar (although not identical) to those of living beings. However, although there are theories about the possibility of building self-reproducing machines, this biological function remains, for the moment, exclusive of living beings. Therefore, reproduction tends to be considered the touchstone that makes it possible to classify any being whatever as living or inert.

It seems, therefore, that a molecule of nucleic acid, which can reproduce spontaneously, should be considered as a living being, even when it is isolated. In fact, several biologists defend this point of view. The American biologist Hermann Joseph Muller (1890-1967), Nobel Prize in Medicine in 1946 for his studies on the genetic effects of X-rays, asserted in the sixties that life has been synthesized in the laboratory, for in 1955 the Spaniard biologist Severo Ochoa built a non-natural molecule of ribonucleic acid, by assembling it from simpler components (nucleotides). This experiment earned him four years later the Nobel Prize in Medicine and Physiology[12].

We can, therefore, consider that nucleic acids are the simplest and most elementary living beings, this stage of evolution being the first and oldest level of life, which probably appeared on Earth about 4 billion years ago.

* * *

Are there now on our planet living beings formed exclusively by molecules of nucleic acid, capable of reproducing and living totally or partially isolated? Perhaps the answer to this question may not be affirmative, but at least there are three groups of beings approaching that concept: viruses, plasmids and viroids. It could be surmised that the first living individuals that appeared on Earth could have belonged to the category of isolated nucleic acids.

---

[12] Other biologists, such as Maynard Smith, consider that reproduction is not enough to define a living being; metabolism should be added. Under this definition, nucleic acids wouldn't be alive. In this book we'll apply Muller's definition.

Viruses were discovered at the end of the last century in relation to certain diseases whose germs, much smaller than bacteria, managed to pass through the finest filters then available. They were called, therefore, filterable viruses[13]. Throughout the twentieth century, several diseases caused by viruses were discovered, both in man, in animals and in plants. Some of these diseases are common and relatively benign, such as the flu, the cold, measles and chickenpox, while others are terrible: polio, smallpox (now eradicated), hydrophobia and AIDS, for instance.

In 1935, the American biochemist Wendell Meredith Stanley (1946 Nobel Prize) isolated in a pure state a virus that causes in the tobacco plant a disease called mosaic. He was surprised when he found that the concentrated virus formed crystalline structures, as if it were a common chemical. For some time there was discussion on whether viruses should be considered as living beings, since the coexistence of the crystalline state with life seemed incompatible.

Viruses are very small living beings. They contain a single DNA or RNA molecule, enclosed inside a protein capsule, sometimes with fats and carbohydrates added. They are incomplete beings, unable to reproduce alone. They parasitize living cells and make use of their machinery to produce their own copies and perpetuate themselves. For this reason, and because their structure is relatively complicated, they are considered today, rather than as true living beings of the first level, as degenerated cells reduced to parasitism.

Plasmids were discovered in 1952 by Joshua Lederberg, who observed that in certain bacteria capable of *conjugating*

---

[13] The Latin word *virus* means poison.

(exchanging genetic information), some genes are transmitted more frequently than others. Further research showed that these genes are found in the bacteria, sometimes isolated, sometimes linked to a chromosome, and that they are capable of reproducing independently and being transmitted by conjugation, even among bacteria of different species.

Plasmids are small molecules of DNA with a few tens of thousands nucleotides, that live inside a bacterium without damaging it, in a mutual agreement or *symbiosis*[14], from which both get benefits. The plasmid can take advantage of the cellular machinery to reproduce, while its genes provide the bacteria with substances that it would have been unable to produce and can confer favorable properties, such as resistance to antibiotics.

Many plasmids are known today, not just in bacteria, but also in cells of multicellular organisms, and sometimes are responsible for important biological processes, such as certain diseases, milk fermentation to produce cheese, degradation of hydrocarbons by bacteria, etc.

The independence of plasmids from the cell housing them is so great that they are sometimes able to survive when their host dies, thanks to the ease with which they can bind to other DNA molecules (including viruses) and be transmitted from cell to cell.

The origin of plasmids is unknown. Perhaps they are residues of an ancient way of life, close to the origin of our planet, consisting of isolated molecules of DNA. However, this is not certain, since they are unable to live and reproduce indefinitely outside the host cells,

---

[14] From the Greek *syn* (together), *bios* (life). Symbiosis means, therefore, *life together*.

so they could be degenerate beings that have moved back to the first level of life from complete cells. Perhaps, when they entered in symbiosis with other living beings, they were forced to renounce all their cellular machinery, with the exception of the essential genetic information that ensures their subsistence.

Finally, *viroids*, discovered around 1962 by Theodor O. Diener and William B. Raymer, are extremely short RNA molecules that belong to a new family of cell parasites, which cause diseases in plants, and possibly in animals and man. The potato viroid, which causes the appearance of fusiform tubers, has just 359 nucleotides in a single chain, sometimes circular, that under certain conditions mates with itself through hydrogen bonds between its bases, to form a linear compact structure. Several viroids are known to attack higher plants, but it is suspected that a being belonging to this group could be responsible for certain animal or human ailments whose cause has not yet been found.

No one knows how such a small RNA molecule is capable of reproducing, since it does not contain enough genetic information to encode a single protein. Until their discovery, it was believed that the minimum genetic mass necessary to ensure the reproductive independence of a parasitic nucleic acid should contain a few thousand nucleotides, as in a virus. But the enigma of viroids, the tiniest living beings, has opened new paths of research on the mechanisms of reproduction of nucleic acids. For instance, an amazing similarity has been observed between viroids and certain strands of DNA that are part of the chromosomes of all living cells. Perhaps viroids have escaped the normal working of cells and adopted an independent and pathogenic lifestyle.

It is possible that neither plasmids nor viroids represent the first life forms on Earth. The living beings of the first level were probably nucleic acid molecules capable of reproducing in freedom, in the absence of cells, by some mechanism now ignored, which may have been abandoned by their descendants, if the course of evolution allowed them to develop more effective methods. In any case, there had to be a time in the history of our planet during which molecules of nucleic acids were the only *living beings* in the waters of the early ocean. From one of those elemental beings may descend all the forms of life inhabiting the Earth today.

3. The second level

The Mesopotamian civilization perhaps knew that transparent objects with curved faces have the property of increasing the size of the bodies seen through them. In the ruins of Nineveh, the sadly famous capital of the Assyrian empire, archeologists found a piece of quartz with a flat and a convex face, which was perhaps used as a magnifying glass. These objects are called lenses[15]. The term was applied to the eye lens by the Greek physician Rufus of Ephesus, towards the end of the first century of our era.

The Chinese civilization also knew about the use of lenses, and applied them to correct eye defects, in the form of glasses. In 1275, Marco Polo witnessed their use in the imperial court. We don't know who introduced them first in Western Europe. At the end of the fifteenth century there were manufacturers of eyeglasses in Germany. The new profession spread to neighboring countries, especially the Netherlands.

At the end of the 16th century, a Dutch eyeglass manufacturer, Zacharias Jansen, placed two lenses at the ends of a long tube, about two meters long, which became the first compound microscope, so named because it had more than one lens, as opposed to the simple microscope, which has one. Thereafter, techniques for the construction of microscopes improved slowly. Around 1665, the English physicist and inventor Robert Hooke (1635-1703) built a better one, by means of which he discovered

---

[15] From the Latin *lens*, meaning *lentil*.

the microscopic structure of cork, and gave the name *cell* to each of the tiny spherical holes it contains. This name, very important in modern biology, comes from the Latin *cellula* (a very small room) and first appeared in Hooke's book *Micrographia. Some Physiological Descriptions of Minute Bodies*, published in 1665.

Shortly after Hooke's discovery, there was a revolution in the techniques of lens carving and microscope construction. The author was the Dutchman Antony van Leeuwenhoek (1632-1723) whose microscopic preparations became famous throughout Europe. Despite his lack of official education, his preparations earned him admission to the English Royal Society, the most important scientific institution of that time, and attracted to his small museum many researchers and celebrities of his time, including members of European royal houses.

Van Leeuwenhoek is famous for having discovered microorganisms: living beings of tiny size, whose existence had been totally unknown before that time. Among those he described are man's sperm and red blood cells; yeasts and infusoria; and some bacteria. The first two belong to higher organisms, made up of billions of cells; the others are complete living beings, able to nourish, to interact with the environment, to reproduce and to make independent lives.

Unicellular organisms (formed by a single cell) are clearly different from the isolated nucleic acids discussed in the previous chapter under the heading *the first level of life*. But not all cells are equal. There are two main classes, very easy to differentiate. Some, simpler, are called *prokaryotes*[16], others, more complex, are called

---

[16] From the Greek *pro* (before), *karyon* (nucleus). So *prokaryote* means

eukaryotes[17]. We will first talk about prokaryotic cells, which, as their name implies, are the oldest. Later we'll deal with eukaryotes.

A prokaryotic cell is made up of a gelatinous liquid body called a protoplasm[18], surrounded by a lipid-protein membrane that isolates it from the outside environment. The protoplasm contains many dissolved and suspended substances, as well as a number of cellular organs that perform specific missions. Among the molecules contained in the protoplasm are all the fundamental principles of living matter: carbohydrates; lipids, especially fats; and proteins. There are also various nucleic acids in their two forms: DNA and RNA. The double helix of deoxyribonucleic acids contains the genetic information of the organism, directs the cellular machinery and supervises reproduction, while ribonucleic acids play the role of messengers between giant DNA molecules and certain cellular organelles (ribosomes) responsible for the synthesis of proteins.

Proteins have a twofold mission: some contribute to shaping the cellular structure, in the membrane or inside the protoplasm; others are enzymes, catalysts that help the cell to carry out chemical reactions, from the synthesis of complex organic products, to respiration.

A single-celled prokaryote behaves as an independent and unique individual despite the fact that many nucleic acid molecules of every kind coexist inside its body. Remember that in the previous

---

before the nucleus.
[17] From the Greek *eu* (true). So *eukaryote* applies to cells with a *true nucleus.*
[18] From the Greek *protos* (first), *plasma* (molded form): *the first molded form.*

chapter we had allowed these nucleic acids the property of being alive, as they are capable of reproduction. A prokaryotic cell is, therefore, a super-organism, a second level being, which contains within itself several living beings of the first level, who act in a coordinated manner, abandoning their individuality in favor of the higher order being made by all of them. Let's see how this integration of efforts takes place.

The physical structure and the behavior of the cell depend mainly on the proteins it contains. These are complex organic molecules built by the assembly of several hundreds of small units called *amino-acids*. The parallelism of proteins with nucleic acids is obvious, as the latter are also made up of chains of simpler elements, the nucleotides. The living beings in the second level have taken advantage of this parallelism to codify the amino-acid composition of proteins by means of the nucleic acid base sequence. Thus, the composition of all the proteins that a living cell can make is encoded in one or a few DNA molecules, capable of reproducing. The codification system used by cells is called *the genetic code*.

There is, however, a difference between proteins and nucleic acids: the latter are made from four different nucleotides, while the amino-acids in proteins can belong to twenty different types. It is clear that the amino-acid sequence of a protein cannot be represented by the same number of nucleotides.

The problem to be solved can be expressed thus: how can we represent twenty different objects by means of a code made by four different letters? For the reader who understands mathematics the solution is obvious. For others, we'll get there gradually. Note, first, that two nucleotides per amino-acid are not enough. If we represent each nucleotide in DNA by its base: A (adenine), G (guanine), C (cytosine) and T (thymine), it can be seen that there are sixteen different combinations of nucleotide pairs, namely:

AA, AG, AC, AT, GA, GG, GC, GT, CA, CG, CC, CT, TA, TG, TC, TT

which is not enough, for there are twenty amino-acids. With three nucleotides, however, 64 different combinations can be formed, which is more than enough. In fact, there are 44 too many. Can they be used? Yes. In fact, no combination of three nucleotides is meaningless. This means that some amino acids can be represented by different triplets: the genetic code is redundant. In addition, there are special triplets that don't represent one amino-acid, but just indicate the end of the nucleotide chain that represents a protein. In this way, a single huge DNA molecule can contain the encoded information of the structure of hundreds or thousands proteins. Each DNA substring that represents a particular protein is called a *gene*.

Table 3.1 shows the genetic code of prokaryotic cells, indicated in RNA nucleotides, rather than DNA. As we know, in RNA, uracil (represented by a U) replaces thymine (T).

Amino-acid	Nucleotide triplets
Phenylalanine	UUC, UUU
Leucine	UUA, UUG, CUA, CUG, CUC, CUU
Serine	UCA, UCG, UCC, UCU, AGC, AGU
Tyrosine	UAC, UAU
Cysteine	UGC, UGU
Tryptophan	UGG
Proline	CCA, CCG, CCC, CCU
Histidine	CAC, CAU
Glutamine	CAA, CAG
Arginine	CGA, CGG, CGC, CGU, AGA, AGG
Isoleucine	AUA, AUC, AUU
Methionine	AUG
Threonine	ACA, ACG, ACC, ACU
Asparagine	AAC, AAU
Lysine	AAA, AAG
Valine	GUA, GUG, GUC, GUU
Alanine	GCA, GCG, GCC, GCU
Aspartic acid	GAC, GAU
Glutamic acid	GAA, GAG
Glycine	GGA, GGG, GGC, GGU
Chain end	UAA, UAG, UGA

Table 3.1. Genetic code of prokaryotic cells

It will be noted that some amino acids are represented by a single triplet; others by two, three, four and up to six different codes.

Each of the three nucleotide groups that encode an amino-acid is called a *codon*.

* * *

The synthesis of a protein is a complex process that goes through several phases. Normally the gene is inactive, electrostatically bound with a special protein (called a *repressor*) that envelops it and prevents its working. At a certain moment, when the corresponding protein becomes necessary, another substance appears that chemically binds onto the repressor that inactivates the gene, and moves it away. When the DNA chain is naked (the gene is said to be active), the double helix separates in that area, as the hydrogen bridge bonds are broken. One of the two copies of the gene then generates an RNA copy of itself.

The new RNA molecule (*messenger RNA*, mRNA in short) is much smaller than the DNA molecule, as it contains the information corresponding to a single gene, while DNA contains many genes. The production of messenger RNA ends in one of the codons that indicate the end of the gene. Since mRNA corresponds to a single protein, it no longer needs special indicators and just contains information about the order of amino-acids.

Once the mRNA synthesis is complete, it separates from the DNA and moves to a cell corpuscle called a *ribosome*. Then a third type of nucleic acid comes into play: a number of RNA chains, much shorter than the previous ones, called *transfer RNA* (t-RNA). Inside all cells there are 61 types of these molecules, each of which contains at one of its ends an *anticodon*, a sub-chain of three nucleotides, complementary to one of the codons that encode an amino acid. At the same time, the other end of the molecule can bind, through a covalent chemical bond, precisely with that amino-acid.

Inside the ribosome, the anticodon of a t-RNA molecule links, through hydrogen bonds, with its complementary codon in mRNA. The amino-acids linked to two consecutive chains of t-RNA are combined with each other and separated from t-RNA. As the mRNA crosses the ribosome, like a thread threaded into a needle, the amino-acid chain grows. When mRNA has completely crossed the ribosome, the protein has been built with the correct amino acid sequence.

We have seen that the synthesis of proteins involves three nucleic acids: the huge DNA molecules, which compact the genetic information of the cell; medium-sized messenger RNA molecules, which direct the synthesis of a single protein; and the small transfer RNA molecules, which control the assembly of each amino acid. In addition, within the ribosomes there are additional RNA molecules, involved in the assembly.

It seems that the prokaryotic cell did not take long to appear, once chemical evolution gave rise to the genesis of complex molecules: proteins and nucleic acids. However, the origin of the first living cells is unclear. Some biologists believe that a parallel evolution, affecting proteins and nucleic acids separately, must have taken place, which could be relatively advanced before both processes were assembled together.

It is likely that the oldest living cells were not very different from the simplest cells we know. The smallest organisms capable of independent living without parasitizing other cells (as is the case with viruses) are *mycoplasmas*, which cause diseases in some domestic animals and possibly also in humans. Some of these cells measure one ten thousandth of a millimeter and weigh less than one billionth of a gram. Their structure is simplified to such an extent, that nucleic acids make over 10 percent of the total weight of a mycoplasma. On the other hand, our classification systems divide prokaryotes into two large groups, *archaea* and *bacteria*, which must have been separated very early in life's history.

* * *

More than three billion years ago, Earth's atmosphere was mainly composed of molecular nitrogen and carbon dioxide. The oldest known fossils are from that era. They are petrified structures, visible only under a microscope, that recall the appearance of certain bacteria. There is no certainty that they are the remains of living beings, but everything seems to indicate that the first cells must have originated at least 3.5 billion years ago, if not farther back.

During the first 2.5 billion years of the history of second level living beings, two very important revolutions happened on Earth. In the first one, photosynthesis appeared, the ability to produce organic matter from inorganic substances by means of solar energy. The first who managed to carry out these processes got a clear advantage against their competitors, who had to find organic food. Therefore, autotrophic[19] organisms multiplied enormously and diversified into many species. On the other hand, heterotrophic[20] organisms were divided into two large groups: predators, which feed on other living beings (autotrophic or heterotrophic), and

---

[19] From the Greek *autos* (self), *trofos* (food); i.e. those who feed by themselves.
[20] From the Greek *heteros* (another); i.e. those who feed through another.

saprophytes[21], that feed on free organic matter, generated by the action of ultraviolet light, or from the remains of dead and decaying organisms.

The first photosynthetic bacteria did not perform the process of organic synthesis through the chemical reactions we see today in green plants. They resembled two groups of current bacteria: green sulfobacteria, and chromatiaceae or purple sulfobacteria, which are capable of performing a special kind of photosynthesis, which uses substances such as hydrogen sulfide (SH_2), sulfur or molecular hydrogen. These bacteria, which today live in oxygen-deprived environments, must have been very abundant at the beginning of life's history, when oxygen did not exist in the Earth's atmosphere.

At some point, long before two billion years ago, a new type of photosynthesis emerged, much more efficient than the previous ones. It uses water as a source of hydrogen, and atmospheric carbon dioxide as a source of carbon. With few differences, this is the method currently used by green plants. These processes were invented by prokaryotic organisms of a new type, which we now call cyanobacteria. Their photosynthesis presented an important novelty: as a result of the set of chemical reactions they use, oxygen is released. In fact, the entire process can be reduced to a very simple equation:

```
Solar light + Carbon dioxide + Water ->
           Oxygen + Glucose
```

This equation is just a summary. Things happen, in reality, in a much more complicated way, but we won't go in so much detail.

---

[21] From the Greek *sapros* (rotten), *fiton* (plant).

As a consequence of these reactions, there came the second great revolution in the history of life. At first, the amount of oxygen in the atmosphere grew slowly. Most of it was absorbed, as soon as it was produced, by iron compounds. Huge deposits of iron oxides (*banded iron* formations) have been discovered, that appear to have been originated at the bottom of the oceans about two billion years ago. But the iron salts dissolved in the sea were soon depleted, and thereafter all the oxygen produced in photosynthesis accumulated in the atmosphere. The equilibrium came when carbon dioxide was reduced to almost zero and oxygen reached its current ratio, one fifth of the volume.

There appeared, in the upper layers of the atmosphere, a special form of oxygen called ozone (O_3), which is opaque to ultraviolet light. This fact, together with the almost total depletion of atmospheric carbon, resulted in the total cessation of spontaneous generation of organic matter. From that moment, the conditions that had made possible the appearance of life disappeared forever, as a result of the action of life.

While the concentration of oxygen was growing, living organisms were subjected to a huge new evolutionary pressure. For most of those living beings then existing, oxygen was a poison, since it is a very active gas, capable of combining with organic substances, destroying them or altering their composition. Prokaryotic cells that lived at that time had to choose between three possible options: extinguish, withdraw to oxygen-free regions (such as the bottom of swamps, where there is a lot of fermenting organic matter), or adapt to the new environment by modifying their structure to resist the presence of oxygen, or even use it for their own purposes. As the increase in the proportion of oxygen must have been very gradual, evolution could act over many generations of microorganisms.

It is almost certain that the three answers happened: some groups became extinct; others, as current sulfobacteria, withdrew to anaerobic[22] environments. Others applied the adage: *if you can't beat it, join it*. Oxygen was tamed, and a new type of prokaryotes appeared, capable of respiration. This was the second revolution.

In the absence of oxygen, living cells extract energy from the fermentation of glucose, getting as end products carbon dioxide plus ethyl alcohol or lactic acid, according to the following summarized equations:

```
Glucose -> Carbon dioxide + Ethyl alcohol + Energy

         Glucose -> Lactic acid + Energy
```

The first is ethyl fermentation, which some yeasts use during the transformation of grape juice into wine. The second is lactic fermentation, which takes place during the transformation of milk into yogurt, as well as in the muscles of mammals during intense exercise, when muscle cells don't receive enough oxygen.

Respiration, on the other hand, is a more complex and efficient process. Its summarized equation is:

```
Glucose + Oxygen -> Carbon dioxide + Water + Energy
```

At first glance it seems that this equation is exactly the inverse of photosynthesis, and so it is, in part. It should be remembered, however, that in both cases they are condensed reactions, which

---

[22] Life in the absence of air (oxygen, in fact).

are carried out through many intermediate steps, which are quite different for the two processes.

A living being extracts a little more than four kilocalories from the total combustion of a gram of glucose, according to the previous equation. However, only 225 calories per gram can be obtained from the fermentation reactions. Respiration is eighteen times more efficient than fermentation. This explains that living beings capable of respiration had an evolutionary success far superior to that of anaerobes. Both their number and their diversification grew enormously; they occupied many ecological niches where their ancestors had not been able to go, and spread across the entire surface of our planet. The oxygen released as a byproduct of vital reactions thus became the key that opened the entrance to a much wider field of evolution, where new species had access to greater amounts of energy. Without this revolution, the Earth would be now populated solely by bacteria.

The transition from fermentation to respiration took place in almost all groups of living beings that existed two billion years ago, among photosynthetic autotrophs, predators and fermenters. There were also some microorganisms that acquired the ability to breathe without losing the ability to survive in the absence of oxygen, thus getting the best out of both worlds, which facilitated their survival in very varied conditions.

Manuel Alfonseca

4. The third level

About fifteen hundred million years ago took place the third revolution in the history of life. A new type of cells emerged, larger and more complex than prokaryotes, whose features gave them a big evolutionary advantage. These are the eukaryotic cells, which obviously came from some aerobic prokaryotic, as practically all of them breathe oxygen. However, to perform this function, they have inside them quite specialized cellular organelles: the mitochondria.

These are elongated corpuscles, about a micrometer in length, with a complex structure, which contain circular DNA molecules and have their own enzyme and protein synthesis system, independent of the one used by the rest of the cell. It is believed that the ancestors of mitochondria were isolated cells that entered *symbiosis*[23] with eukaryotes at the beginning of their history and became a part of them. A surprising discovery has come in support of this theory. It has been possible to prove that the genetic code of prokaryotes (the one in table 3.1), which at first was considered universal, does not apply to mitochondria, as there are small discrepancies: in these organelles, the UGA codon represents amino-acid Tryptophan, rather than indicating the end of the gene, while the AUA codon corresponds to methionine, rather than isoleucine. It seems that mitochondria originally belonged to the

---

[23] See note 14.

archaea group, while the host organism was an ordinary bacterium. If this interpretation is true, eukaryotic cells can be considered as individuals belonging to a new level of life: the third.

The most important difference between prokaryotic and eukaryotic organisms is precisely what their name indicates. Eukaryotes have their genetic material (DNA) organized in the form of chromosomes contained inside a special organelle, the *nucleus*, which is provided with a membrane that separates it from the rest of the protoplasm (called *cytoplasm*).

This organization is paired with a very complicated cellular reproduction system, *mitosis*, which opens the way to sexual reproduction. Prokaryotic organisms, on the other hand, reproduce by bipartition or division into two equal parts of a single mother cell, although they also have mechanisms that allow them to exchange genetic information, thus achieving, to some extent, some of the advantages of sexual reproduction. These are very important, for if a living being has only one parent, it may inherit the beneficial genes, but it must also take the harmful. On the other hand, if every individual has two parents, the shuffling of genetic information that takes place during fertilization leaves a certain probability that the favorable genes of both parents will be combined and the unfavorable ones avoided, so that natural selection will act in favor of that individual, and evolution accelerates. But this will be discussed in more detail in the next chapter.

Some eukaryotes are able to perform photosynthesis in the style of cyanobacteria, although, as with respiration, they contain special organelles, called *chloroplasts*, which also appear to be old isolated cells, which today live in a state of symbiosis, such as mitochondria. Other eukaryotes are heterotrophs and must feed from organic substances manufactured by other organisms. In the classification system, unicellular eukaryotes are divided into several groups, which can be grouped under the names of unicellular algae, unicellular fungi and protozoa. Only the former are autotrophic and have chloroplasts. The other two groups live a saprophytic or a predatory existence on second and third level organisms, or are parasites of fourth level organisms.

Manuel Alfonseca

5. The fourth level

The existence of the living beings in the first level has been discovered over the last hundred years. The cells of the second and third level, capable of subsisting independently, were totally unknown until about three hundred years ago. However, plants and animals were well-known to man since his origins, about two million years ago.

The reason is evident: we are, like plants and animals, living beings of the fourth level, we feel that we are individuals, and find it easy to attribute equivalent properties to similar entities. The difficulty grows as we move away from our degree of evolution in any direction: it's not difficult to accept that a cell is alive, but the application of the same quality to a DNA molecule encounters resistance. Something similar happens in the opposite direction, as we'll see in successive chapters, when we deal with the fifth level. Everything happens as if we, human beings, had stepped on the fourth step of an ascending staircase but, affected by a curious optical effect, we believe that we are at the top, and it seems to us that the ladder descends in all directions.

What is a living being of the fourth level? An individual made of the union of many living individuals of the third level, who give up their independence for the good of the whole they all make together. This implies a set of phenomena similar to those already mentioned when dealing with the steps between the three previous levels. The most significant is the great diversification of the members of the lower level when thy join and make a more complex individual: in the same way that there are many different types of nucleic acids inside a prokaryotic cell (chromosomal DNA, messenger RNA, transfer RNAs, ribosomal RNA, etc.), and there are several different types of prokaryotic cells within an eukaryotic cell, there are also cells of many kinds in the body of a plant or an animal: cells that differ more from one another by their shape or properties, than any pair of prokaryotes or eukaryotes making independent life, whatever their species.

Although the cells that are part of an individual of the fourth level can have very different shapes, diversity is not chaotic. The different types of cells complement each other to form a coordinated harmonic whole, as was the case with the various nucleic acid species inside a living cell. The cells of the body of a plant or an animal can be classified into a not-too-large number of different types: they form families of cells that live together and build organized tissues, whose mission in the individual of the fourth level can be very varied: protect the body and the surfaces of important organs (epithelial cells); provide movement (muscle cells); link various organs and tissues with each other (connective cells); defend the organism against the attack of foreign substances and living beings (cells of the immune system); provide support to maintain a constant shape or composition of the body (cartilaginous, bone and woody cells); provide the body with a control system that makes all the parts work together and provides information about the surrounding environment (nerve cells).

In the same way proteins and nucleic acids are assembled in a living cell to form organelles with specific missions (ribosomes, nucleus, etc.), so also the different cells and tissues join together to make organs that let the living being of the fourth level continue living, react to environmental stimuli and propagate its species: the previously mentioned functions of nutrition, relationship and reproduction.

The total number of living individuals of the fourth level is enormous: it is estimated that there are about one quintillion insects (10^{18}). Fortunately, since ancient times, man has noticed that many of these beings have common properties; that they resemble each other. These properties, called universals, make it possible to make classifications and distribute all those individuals in a much smaller number of groups, making possible language and communication between human beings. If we were to give a different name to each thing we see, we couldn't agree on what name to use for each one. Thanks to universals, the word *cat*, for example, does not refer to a single entity, but to a whole class of beings. In this way, a single name can be used to designate millions of objects.

The basic category used to classify the living beings of the fourth level is the *species*, which is defined as a group of individuals closely linked by kinship relationships and capable of reproducing among themselves, producing fertile issue. Any two individuals of a given species always have common ancestors whose appearance was not very different from theirs. The bad news is that the total number of known species is huge. In the case of animals, it exceeds one million, and it is estimated that there are about 400,000 species of fungi and vegetables. If we add still undiscovered species, we are possibly talking about 30 million species on our planet.

The first person who tried to classify living beings was the Greek philosopher Aristotle (384-322 BC), who devised the famous division of animals into vertebrates and invertebrates, now abandoned in scientific classifications. However, Aristotle's zoological system recognized just fifty species. His disciple and successor as president of the Lyceum, Theophrastus of Eresos (372-287 BC), continued along the same lines and undertook the systematization of the vegetable kingdom. As a result of his efforts he wrote two works: *Enquiry into plants* and *On the causes of plants*, where he lists about five hundred different species of plants.

Modern systematic biology, which deals with the classification of living beings, divides those of the fourth level (also called multicellular[24]) in three great kingdoms: Fungi; Metaphyta[25] or plants; and Metazoans or animals. Each of them is subdivided, in turn, into lower taxonomic categories. The main ones are called *phylum*[26], class, order, family, genus and species. The existence of these categories makes it possible to express with great simplicity the degree of kinship between two living individuals of the fourth level.

The classification of multicellular beings, according to this system of categories, can be represented as an inverted tree: the taxonomic tree. The successive levels of the tree correspond to the taxonomic categories. All individuals who live today or existed in the past are at the lower level. An important advantage of this tree is that it makes it possible to segment systematic biology into simpler

---

[24] From the Latin *multus*, many: consisting of many cells.
[25] From the Greek *meta*, with, together with; *fyton*, plant.
[26] A Greek word meaning *branch*.

branches (zoology, botany, entomology, ornithology, etc.), each of which can be studied as an independent field, while keeping uniformity of structure.

The degree of kinship between two given individuals can be measured by the level of the lowest taxonomic category including both of them. The relationship will be maximum if both belong to the same species, the minimum taxonomic category. On the other hand, a plant and an animal, which belong to different kingdoms and, therefore, have as their only common category the root of the tree (the fact of being multicellular) have the farthest possible kinship.

* * *

The modern concept of biological species was first introduced in the 17th century by the English naturalist John Ray (or Wray, 1627-1705), who is considered the father of natural history in the United Kingdom. Less than a century later, the Swedish Karl von Linné (Carolus Linnaeus, 1707-1778), laid the foundations of modern taxonomy in his book *System of Nature*[27] and other works, where he introduced the binomial nomenclature, which assigns each biological species a double name, as is done with human beings. The first part of the name, called genus, corresponds to our family name and is capitalized. The second, the specific name (the name of the species), is written in lower case and corresponds to our first name. In this way, the kinship of the species belonging to the same genus is recognized immediately, since they all share the same family name, the same genus.

---

[27] *Systema Naturae*, 1735.

At first it was believed that biological species were fixed and immutable entities. In the words of Linnaeus himself: *There are as many species as diverse forms produced in the beginning by the Infinite Being; whose forms have produced others, always similar to themselves, according to the laws established for generation*[28].

This point of view received its first blow when it was shown that fossils[29] were the remains of long dead animals or plants, whose hard parts have been subjected to chemical and geological processes that have turned them into stone. Fossils were known in ancient times: the discovery of petrified seashells in mountains moved Xenophanes of Colophon (6th century B.C.) to assert that these regions may have been in remote times under the sea.

During the Middle Ages it was thought that fossils were vagaries of Nature, stones that had taken shapes resembling those of animals and plants, but did not have an organic origin. Leonardo da Vinci (1452-1519) suggested that they could be petrified remains of animals and plants. The idea was accepted, but for a long time it was assumed that they belonged to the same species that we can see today.

At the end of the eighteenth century, the Frenchman Georges Cuvier (1769-1832) proved by means of new techniques in comparative anatomy that some fossil elephants were not identical to those existing today, and should be considered extinct. It did not take much time for the first paleontologists to see that most fossils correspond to beings that have disappeared from the Earth. It was thus possible to verify that the faunas and floras of past times

---

[28] *Philosophiae Botanica*, 1751.
[29] From the Latin *fossio*, digging. A fossil is, therefore, what is got by digging.

differed a lot from the current ones and, although there are a few common forms, they are relatively rare.

The idea that species evolve, transform and descend from each other, was latent and sooner or later had to come to someone's mind. The first to formulate a theory of evolution to explain the origin of species was the Frenchman Jean Baptiste Pierre Antoine de Monet, Chevalier de Lamarck (1744-1829). In his book *Zoological Philosophy*[30], he argued that those features acquired by an individual in response to environmental pressure can be transmitted to its descendants. The progressive accumulation of these characters would lead to the drift of species and cause their gradual evolution.

Charles Robert Darwin (1809-1882) took the decisive step. On a trip around the world from 1831 to 1836, he had the opportunity to study life forms and fossils of several continents. In particular, a group of finch species found in the Galapagos Islands, where each island has its characteristic species, which have occupied diverse ecological niches that, in other regions, correspond to birds of different families. These facts were the seed that developed in Darwin's mind until it became the theory of evolution.

Another fundamental source for his ideas was the essay by the English economist Thomas Robert Malthus (1766-1834), *Essay on Population*, from which Darwin extracted the principle of natural selection, which he then applied to living species. But it took more than twenty years before he decided to publish his work, which he finally did because another man had independently reached the same conclusions, starting from similar sources.

---

[30] *Philosophie Zoologique*, 1809.

Alfred Russell Wallace (1823-1913) had also read Malthus's essay. Like Darwin, he traveled all over the world on scientific, botanical and zoological expeditions. In 1858 he wrote a work entitled *On the tendency of varieties to depart indefinitely from the original type*, and sent to Darwin a copy, requesting his opinion.

Darwin then decided to publish, at the same time as Wallace, a communication summarizing his own conclusions. A year later his great work, the book on which he had been working for almost twenty years, was published: *On the origin of species by means of natural selection*.

* * *

What is *natural selection*, on which Darwin based his revolutionary theory of evolution? The individuals of a population are never identical to each other. Some of them may differ in physical strength; protection against enemies; greater speed; resistance to certain diseases... Statistically, those individuals who enjoy some advantageous trait are likely to survive longer, and therefore will have more offspring, than those who lack them or have harmful traits. Over several generations, the descendants of those favored by natural selection will be more numerous than those of their competitors.

But if environmental conditions change (a climate cooling, a progressive desertification, the invasion of the territory by a new predator or a competing species), it may happen that previously favorable traits become harmful. If that occurs, natural selection will act in favor of those previously eliminated and the proportion of the traits will be reversed. In extreme cases, individuals who possess negatively selected traits can be completely extinguished.

The Fifth Level of Evolution

* * *

By the end of the 19th century, evolutionism had been almost universally accepted by modern science. However, there were two important doubts regarding the mechanisms of the evolutionary process. How do differences arise between individuals? How are they transmitted from parents to children? In Darwin's time it was believed that children always inherit intermediate traits from their parents. If that were the case, any new trait, however beneficial, would be drowned by hybridization and mixing in a few generations. Being new, it would obviously be scarce.

The laws of inheritance were discovered by the Austrian Augustinian monk Gregor Mendel (1822-1884). In the garden of his monastery at Königskloster, near Brno (Moravia), he carried out, for several years, cross-breeding experiments of several varieties of peas, which gave surprisingly regular and mathematically simple results.

Mendel, who was unknown in the scientific world, sent a summary of his work to the then famous Swiss botanist Karl Wilhelm von Nägeli (1817-1891), who returned it accompanied by a negative review. In spite of that, Mendel published in 1865 a work entitled *Research on plant hybrids*[31] in the minutes of the Natural History Society of Brünn (Brno). The article went unnoticed and Mendel, who in the meantime had been appointed abbot of his monastery, dedicated himself to other activities.

In 1900, sixteen years after Mendel's death, three botanists from various countries[32] rediscovered independently Mendel's laws. All

---

[31] *Versuche über Pflanzenhybriden.*
[32] Hugo de Vries, Dutch; Karl Correns, German; Erich Tschermak,

three investigated the scientific literature, found the article by the Augustinian monk and publicly acknowledged his priority. Thirty-five years late, the scientific world appreciated the importance of Gregor Mendel's discoveries.

* * *

Mendel's laws answered one of the two outstanding questions about the theory of evolution: the hereditary transmission of different traits. The first one remained to be answered: in which way can new traits arise, which did not exist before, so that they compete with previous ones and, perhaps, replace them? The answer was found by the Dutchman Hugo de Vries (1848-1935), one of those who rediscovered Mendel's laws.

Certain abrupt changes in the characteristics of a few individuals of a species had been known for a long time. Those traits differentiate them from the previous generation. Farmers are used to these developments. A famous case was the sudden appearance of yellow canaries in the 18th century. In 1886, de Vries observed one of these abrupt changes in a plant, *Oenothera lamarckiana*. After many years, he found the explanation. In the work that summarized his research[33], he called them mutations. It is an irony of science that the plant that finally strengthened Darwin's evolutionary theory against Lamarck's theory was named after the latter.

The experiments of the American biologist Thomas Hunt Morgan (1866-1945, Nobel Prize in Physiology and Medicine in 1933), carried out mainly on the fruit fly *Drosophila melanogaster*,

---

Austrian.
[33] *Die Mutationstheorie*, 1901-1903.

developed the nascent science of genetics and eventually led to the proof that mutations are nothing more than modifications in the structure of genes (the segments of DNA that encode proteins in living cells). The cause of a mutation may be one of the following: the exchange of one base for another within a gene, so that the codon to which it belongs will encode a different amino-acid; the loss of a segment of the gene; the insertion of a new section; or the reversal of the order of a sequence of bases.

Mutations can occur in any cell in the body of a plant or animal, but only those that occur in the reproductive cells (*gametes*) have the potential to be transmitted to the offspring. The causes can be quite diverse: radioactivity (including what reaches us through the atmosphere in the form of cosmic rays), high frequency electromagnetic radiation (ultraviolet rays, X-rays and gamma rays), some chemical substances, extreme temperatures, and many others.

* * *

Between 1920 and 1940 all these scientific facts and theories were integrated in a new body called Neo-Darwinism or synthetic theory of evolution. Evolution, according to this theory, does not act on isolated individuals, but on whole populations. The mechanisms that cause it are genetic variability and natural selection. A population consists of a group of individuals, each of whom has a specific genetic endowment, the result of the accumulation of mutations and their transmission according to Mendel's laws. In the total population, many genes are represented by two or more different varieties, called *alleles*. A species doesn't have, therefore, a unique genetic composition, but a mixture of alternatives that, thanks to sexual reproduction, are shuffled from generation to generation. Each alternative can appear in the population with a different frequency. Some alleles will be owned by most individuals, while others would belong just to a few. In other species, the population will be split almost equally between two or three alleles.

Natural selection acts on these alternatives in a purely statistical way. Nothing prevents, in fact, that specific individuals, no matter how gifted, fall prey to a predator as a result of an accident, while another badly gifted individual can escape and reproduce, as a result of a favorable combination of circumstances. But in the long run, in a population that contains thousands of individuals, those best adapted to the environment will be able to reproduce more frequently than the less fit.

Genetic variability makes it possible that many species of living beings won't be extinguished at the smallest change in the surrounding environmental conditions. A gene, which up to a point in time was beneficial, can become abruptly unfavorable when external circumstances have changed. The opposite can also happen: an allele that had been negatively selected may suddenly become favored. Natural selection acts on the genetic variability of the population. The gene that must be favored is usually present in the population, even if only in minimal proportions. It is not necessary to depend on a happy mutation to produce its sudden appearance at the right time.

With the synthetic theory of evolution, Neo-Darwinism was universally accepted by biologists and became one of the foundations of current scientific thinking, almost at the same level as Newton's laws, relativistic mechanics, or atomic theory. This does not mean that there aren't pending questions, or that we have reached the limits of research. Some questions still have no clear answer. For instance: is evolution a consequence only of natural selection, or are there indifferent traits, as asserted by the *neutralist theory of evolution*, poised by Motoo Kimura? Does evolution always act gradually, or sometimes accelerates, while at times species remain stable for long periods of time, as in *punctuated evolution*, by Stephen Jay Gould?

* * *

The remainder of this chapter briefly outlines the processes that led to living beings of the fourth level and their progressive evolution over a billion years.

There are currently some animals called *mesozoa*[34], which appear to be intermediate between eukaryotic beings (third level) and multicellular (fourth level). The mesozoa live as parasites in the kidneys of cephalopods (octopus, squid, etc.), are elongated, a few millimeters in length, and consist of a single axial cell, around which about twenty cells are joined together. Despite their simple structure, they can use sexual reproduction, although they do it only when their population exceeds a certain density.

It may be that the mesozoa are metazoans who have degenerated through parasitic life, or perhaps they truly represent an intermediate stage between the third and fourth levels of life. Be that as it may, the union of independent cells to form unique individuals happened several times about a billion years ago, during a geological era that paleontologists call *Proterozoic*[35].

---

[34] From the Greek *mesos*, intermediate, *zoon*, animal.

Each of the three kingdoms of multicellular beings reached independently the fourth level of life following several ways. It is believed that Metaphyta (plants) crossed the threshold by at least six different paths; Fungi, by four or five; Metazoans (animals) at least twice: sponges on the one hand, all the other groups on the other.

Fossils begin to be abundant about six hundred million years ago, when living beings with hard parts, capable of petrification, begin to proliferate. For this reason, modern paleontological nomenclature gives the last six hundred million years the name of *Phanerozoic* eon[36].

The Phanerozoic is divided into three geological ages and these into two or more periods, according to table 5.1, which also indicates the time, in millions of years, elapsed from the beginning of each of the periods to the present day.

---

[35] From the Greek *proteros*, the first. This is the era of the first animals.
[36] From the Greek *phaneros*, visible: the era of animals made visible through their fossils.

Eon	Era	Period	Beginning (million years)
Cryptozoic	Archaic		4600
	Proterozoic		2500
		Ediacaran	1100
Phanerozoic	Paleozoic	Cambrian	565
		Ordovician	510
		Silurian	440
		Devonian	410
		Lower Carboniferous	365
		Higher Carboniferous	325
		Permian	290
	Mesozoic	Triassic	251
		Jurassic	205
		Cretaceous	135
	Cenozoic	Tertiary	65
		Quaternary	1

Table 5.1. Eons and geologic eras

All the animals and plants of the Cambrian period[37] were exclusively aquatic, so the continents were deserted and devoid of multicellular life. However, the main types of metazoans had already diversified, even though many of the fossils of that time correspond to frankly primitive beings. We have also found remains of animals belonging to types that do not exist today, because they later became extinct.

---

[37] Derived from *Cambria*, Roman name of Wales.

Table 5.2 shows the names of the main metazoan phyla, along with some of the current animals belonging to each. There are now more than twenty phyla, in addition to another eight or ten known only by fossil remains from the Cambrian period.

Phylum	Present animals in the phylum
Porifera	Sponges
Coelenterata	Jellyfish, corals
Platyhelminthes	Taenia and other parasites
Aschelminthes	Rotifers, nematodes, etc.
Mollusca	Snails, clams, cephalopods
Annelida	Leaches, earthworms
Arthropoda	Crustaceans, arachnids, insects
Echinodermata	Sea stars and urchins, Holothuria
Chordata	Fish, amphibians, reptiles, birds, mammals

Table 5.2. Metazoan organization types

Along the Paleozoic era[38] took place the invasion of the mainland, first by plants, then by arthropods, finally by vertebrates (amphibians and reptiles). To be able to conquer the mainland, reptiles relied on a very important innovation: the reptilian egg, a complex structure with insulating shell, food (the yolk sac) and a membrane filled with liquid (the amnion) containing the embryo. The egg thus reproduces the environment where the amphibian larvae develop, but by being isolated, it can be left anywhere.

---

[38] From the Greek *palaios*, old. The era of ancient animals.

The end of the Paleozoic era is marked by the greatest catastrophe in the history of the fourth level of life: 90 percent of the species on solid ground disappeared, three-fourths of amphibian families and four fifths of reptiles. At that time, all the continental blocks joined in a single super-continent (Pangaea[39]), but the cause of the mass extinction was perhaps the impact of an extraterrestrial body (an asteroid or a comet) or a tremendous number of volcanic eruptions.

Be that as it may, the diversity of the living beings of the fourth level was extremely reduced at the beginning of the Mesozoic era[40]. This is the time when birds and mammals appeared, although the era is usually known as *the empire of dinosaurs*[41]. I guess the reader will agree that, of all extinct animals, dinosaurs are the best known and those that make most impression on children and adults of all ages.

At the end of the Mesozoic era, about sixty-five million years ago, a new catastrophe took place. The fauna of oceans and continents was decimated. The flora, on the contrary, did not suffer much damage. The large reptiles disappeared from the Earth, the same as the ammonite cephalopods and a group of protozoan. The extinction was due to the impact of the asteroid that made the Chicxulub crater in Yucatan. This theory was proposed by Luis Walter Alvarez (Nobel Prize in Physics in 1968, for his work on elementary particles).

---

[39] From the Greek *pan*, the whole, *Ge*, the Earth. Pangaea means, literally, *the whole Earth*.
[40] From the Greek *mesos*, intermediate. The era of intermediate animals.
[41] From the Greek *deinos*, terrible. Dinosaurs were, therefore, the *terrible lizards*.

So we enter the Cenozoic era[42] with a relatively unpopulated Earth, as regards large animals. This was the opportunity of mammals, who until then had not been able to exceed the size of a rabbit, for all the ecological niches for larger animals were occupied by reptiles. Among mammals, those called Eutheria or Placental had invented an even better method than the reptilian egg to protect the embryo during the early stages of its development: the fetus remains inside the mother's body, connected through a membrane (the placenta) that provides food and eliminates waste substances. At the beginning of the Tertiary period, the Placental supplanted all the other groups of primitive mammals, which were reduced to isolated areas, such as the Australian continent, or to a few resistant species, such as opossums. The Placental occupied most of the mainland, learned to fly (bats) and returned to the sea (Pinniped, Cetacean and Sirenian).

---

[42] From the Greek *kainos*, new. The era of new animals.

6. What is man?

Let us now focus our attention on one order of placental mammals that bears the honorary name of *primates*, from the Latin *primus*, 'the first one'. Why the first? Certainly not because it's the oldest, but because this is the group including *man, the current summit of evolution.*

At the beginning of the Tertiary period, primates were small animals, similar to insectivores. Closest to them today are *Tupaiidae*, arboreal animals that live in the jungles of Southeast Asia. Lemurs and similar primates came later; today they are reduced to isolated populations in Central Africa, India, Southeast Asia and, especially, the island of Madagascar. An extinct family of primates, the *Adapidae*, that lived about fifty million years ago, seems to provide a link in the chain that led to the higher primates (American monkeys, Old World monkeys and hominoids). Table 6.1 summarizes a classic classification of the Hominoid superfamily, which includes man and anthropoids.

Family	Genus	Lived million years ago
Pongidae	Proconsul	22-14
	Sivapithecus	15-8
	Dryopithecus	10-9
	Oreopithecus	7
	Gigantopithecus	6-0,3
	Pongo (orangutan)	Now
	Pan (chimpanzee)	Now
	Gorilla (gorilla)	Now
Hominidae	Ardipithecus	7-4,4
	Australopithecus	4-1
	Homo (man)	2,6-Now

Table 6.1. Classification of superfamily Hominoidea

It seems that hominoids separated from Old World monkeys twenty-five or thirty million years ago. Later, the ancestral type of hominoid, which perhaps resembled Proconsul, resulted in several different evolutionary lines, classified into the two families of the *Pongidae* (anthropoid apes) and hominids (man and his direct ancestors).

The classification of hominids has been subject to curious oscillations. At first, each skeleton found was classified as a new species, even a different genus: *Pithecanthropus, Sinanthropus, Australopithecus, Paranthropus, Meganthropus, Zinjanthropus, Homo heidelbergensis, Homo neanderthalensis,* etc. This was due to the humanly understandable wish of researchers to put important discoveries under their name. It is easy to see that finding a new bone of *Homo erectus* is less spectacular than discovering a new species or a new genus.

In the mid-twentieth century, faced with the tremendous proliferation of scientific names of the ancestors of man, paleontologists decided to put a little order and reduced all existing specimens to two genera (Australopithecus and Homo) and seven species (four of the first, three of the second).

Since the seventies to the present, new discoveries have been made, and the primitive tendency has prevailed again. When the first hominid remains older than *Australopithecus* were found, new genera began to proliferate, such as *Ardipithecus*, *Sahelanthropus*, *Orrorin*, *Keniapithecus*, as well as new species of the genus *Homo*, such as *Homo rudolfensis*, *Homo ergaster* or *Homo antecessor*. However, since the beginning of the 21st century, there are signs of a new simplification, in the form of proposals to return to the three classic species of the genus *Homo*, or to group all the hominid genera that lived 7 to 4 million years in a single genus: *Ardipithecus*. This is the version we are adopting here.

The first hominids appear to have separated from the line that led to the chimpanzee 5 to 7 million years ago. We classify them in the genus *Ardipithecus*, which may have consisted of three or four species. Four million years ago, appeared the second hominid genus, *Australopithecus*, which currently consists of four species (*A. afarensis*, *A. africanus*, *A. robustus* and *A. boisei*, the last two sometimes called *Paranthropus*). They lived since 4 million years ago, until just under a million. These beings walked upright and had a brain capacity less than half a liter. Finally, the third genus appeared, where man is classified.

Three successive stages of the genus *Homo* are known, differing both in their anatomical characteristics and in their culture (the stone instruments they made), which usually receive the range of biological species, although there is controversy in this respect. The first, *Homo habilis*, lived in Africa 2.6 to 1.5 million years ago. He had a brain about eight hundred cubic centimeters and was able to use very simple tools: he is the author of the pebble culture (pebbles sharpened by blows) found in some African sites.

The oldest remains of *Homo erectus*, the second stage of the genus *Homo*, come from Africa and date from 1.9 million years ago. The most modern go back to a little more than one hundred thousand years ago. This is a more advanced type of man, who knew how to make fire, carved stones to make sharp instruments (he is the author of the *Acheulean* culture) and had a brain capacity of nine hundred to twelve hundred cubic centimeters (about one liter). Fossils of this species have been found in Africa, Asia and Europe; thus *Homo erectus* managed to cover the whole Old World.

With *Homo erectus* we have crossed the border that separates the two periods in which the Cenozoic era is divided: we are now in the Quaternary. At that time, the weather underwent significant modifications: several glacial epochs, separated by intermediate milder periods. Two or three hundred thousand years ago appeared the third stage of genus *Homo*: modern man, with a brain capacity of twelve to sixteen hundred cubic centimeters, which Linnaeus called *Homo sapiens* (wise man). Today Neanderthal forms and those similar to present man are usually classified under this name.

* * *

The Fifth Level of Evolution

Let the reader get back for a moment to the first paragraph in this chapter. Perhaps you were surprised, the first time you read it, by the last sentence, the one written in italics: *man, the current summit of evolution*. This phrase wouldn't have surprised anyone in the first century after Darwin: everyone took it for granted, considered it obvious. What has happened during the second half of the twentieth century, so that it is no longer obvious?

Quite simply, that the dignity of man has suffered its greatest attack in history. It is curious that this has happened precisely at the same time that the United Nations approved in 1948 the Universal Declaration of Human Rights. The attack comes from materialistic atheism, which has spread through the media, and whose purpose is the debasement of man to the level of just one animal, to deny our possible transcendence (the immortality of the soul, in classical language).

The idea that biological species are not at the same level; that some can be considered higher to others, is now almost considered anathema. After the fundamental equality of human beings, usually attributed to the French Revolution (although St. Paul had said it before), we are moving on to the fundamental equality of living species: a louse is not lower than man, please!

In the words of biologist Julian Huxley[43] (1887-1975): *Man's opinion of his own position in relation to the rest of the animals has swung pendulum-wise between too great and too little a conceit of himself... The gap between man and animals was here reduced not by exaggerating the human qualities of animals, but*

---

[43] Biologist, grandson of Darwin's famous disciple, T.H. Huxley, and brother of Aldous Huxley, the writer, and of Andrew Fielding Huxley, who won in 1974 the Nobel Price in Physiology and Medicine.

by minimizing the human qualities of men[44]. Another famous biologist, George Gaylord Simpson (1902-1984) says: *[man] is another species of animal, but not just another animal. He is unique in peculiar and extraordinarily significant ways*[45]. Finally, Theodosius Dobzhansky (1900-1975) wrote about this: *[Classical evolutionism] emphasized the many respects in which human beings are fundamentally similar to other biological species. At present, it is more important to study in what ways they are unlike other species.* Later he adds: *The human species is unique in the living world because of a complex of interdependent characteristics. Although some [of them] may be found as rudiments in non-human animals, as a complex they are the property of humankind alone... [Thanks to culture] human evolution has transcended, that is, it has gone beyond the limits of biological evolution*[46].

Cladistics is a modern version of systematics, the part of biology that deals with the classification of living beings. This discipline tries to order all living beings in a more natural way than traditional classifications by building the family tree of the species, so that two species that have separated more recently must be closer to each other than any other two that separated before.

In a book describing the methods and results of cladistics, Colin Tudge[47] considers this science as a revolution similar to that initiated by Copernicus, because it has removed *Homo sapiens*

---

[44] *Man stands alone*, 1941.
[45] *This view of life*, 1964.
[46] *Human culture: a moment in evolution*, posthumous work finished by Ernest Boesiger, 1983.
[47] *The variety of life*, Oxford University Press, 2000.

from the supreme position in nature, by increasing the number of kingdoms in our classifications. Before cladistics we had just two; animals and plants; now we have more than fifty, because most groups of unicellular beings, what we have called here the second and third levels of life, now receive the rank of kingdoms. So we are not important: after all, we are just members of a kingdom among fifty, while before cladistics we were the members of a kingdom among two.

The argument is amazing. It could serve to prove that Rembrandt cannot be considered a great painter: let us build a genealogic tree of painting based on cladistics (which is easy), and let's call every primitive painter a school or a civilization, while modern masters become just members of lower level groups.

In fact, importance has nothing to do with the number of schools or kingdoms, or with the order of their appearance. Cladistics gives us useful information about the origin of species over time, but a modern classification system should take into account other things, such as complexity. The four levels of living beings, described in the previous chapters of this book, give us a scale to measure complexity, much more reliable than the fifty kingdoms of cladistics.

Since the eighteenth century, when *the myth of indefinite progress* arose, we are used to despise ancient and medieval scientific knowledge, accusing our ancestors of ignorance and preferring myths to the hard facts on which science is based. If the issue is studied carefully, one must conclude that we are often carried away by appearances and encourage the proliferation of pseudoscientific myths and false knowledge, which attract the attention of mass media and spread quickly, becoming almost ineradicable.

I will mention a couple of examples: we've heard frequently that *in ancient times and in the Middle Ages everybody believed that the Earth is flat*. Well-informed people know that this commonplace is false, but it is fully believed by the man in the street. The truth is different: twenty five hundred years ago, the Greeks knew that the Earth is a sphere (Aristotle mentions three independent demonstrations), and during the Middle Ages, only the ignorant believed that the Earth is flat, and that the ships would fall if they reached *the end of the world*.

It is also common to hear that *in ancient times and in the Middle Ages it was believed that the Earth is very large: modern astronomy has shown that it is infinitesimal*, compared to the universe. This myth is even more widespread than the previous one, but it is equally false. Archimedes calculated that the radius of the Earth is at least one billion times smaller than the distance to the nearest star, a correct order of magnitude, while Claudius Ptolemy (whose *Almagest* was the standard astronomy text throughout the Middle Ages) wrote: *The Earth, in relation to the distance of the fixed stars, has no appreciable size and should be considered as a mathematical point* (Book I, Chapter 5).

A different myth, which concerns us here, was derived from the previous one: man used to think he was the most important object in the universe, but modern science has shown that we are really unimportant. First, Copernicus took our planet out of the center of the universe; then Darwin and cladistics proved that we are just one species among many; the study of the human genome shows that we are practically chimpanzees and little more than flies; modern physics, that our body is made up of insignificant atoms. The first two statements are in fact quite old, the last two are recent.

It is false that man was considered during the Middle Ages the most important being in the universe. In the Divine Comedy, Dante, following Ptolemy's cosmology, travels through the celestial spheres. Upon reaching Saturn, he looks back at the Earth, which seems really tiny. In consequence, he finds contemptible most of the problems that usually concern man (*Paradiso*, 22:133 and following).

The latest discoveries about the genome are always presented in the media as humiliations that we must suffer, which should lower our dignity. In fact, they are simple numerical findings that don't have such consequences. It is said, for example, that the human genome matches the genome of the chimpanzee by 98.5%. From this, they try to deduce that we are practically identical to the chimpanzee. The conclusion is fallacious, as I will prove with a simple example: according to the same criterion, we might think that pure water at normal pressure and 273°K must be practically identical to water at 274°K. After all, the two temperatures only differ by 0.36%, much less than the genomes of man and chimpanzee. However, in fact, they are totally different: the first is solid (ice) and the second is liquid.

This example teaches us that the world is not linear, that there are abrupt growths, stagnations and sudden changes of state, such as what takes place at the melting and boiling points of water. Data on the genomes of man and chimpanzee just show that a difference as small as 1.5% was enough to cross the threshold of reason, which has placed us on a completely different level. On the other hand, these calculations are not always statistically valid, since they are not usually performed on the complete genome of man and chimpanzee, just on a sample that rarely exceeds 1% of the genes. Other results, obtained by comparing proteins, seem to give quite different figures.

In an article published in Nature on May 27 2004, several researchers announce that they have compared the functional genes of a human chromosome with that of the chimpanzee. Their results are surprising: as expected, 98.5 of the nucleotides coincided, but of 231 functional genes compared, 47 differ and produce different proteins. In other words, 1.5% differences in nucleotides are distributed throughout many genes, leading to a protein differentiation greater than 20%. However, this news has not received the same attention by the media. By increasing the genetic difference between man and chimpanzee, it takes us away from animals, so it's better to hide it. I have no choice but accuse the media of lack of professional honesty, since they only publish the news that favors their ideology.

Let's move on to the second statement. In an article published in 2000, Michael S. Turner[48] discusses the state of research on the mass of the cosmos (at that time), classified as follows: 65% dark energy (which would cause the accelerated expansion of the universe); 30% dark matter (made of unknown particles); 4% hydrogen and helium, dispersed in galaxy halos; 0.5% neutrinos; 0.5% condensed matter in the form of visible stars and galaxies. Of

---

[48] *More than meets the eye*, The Sciences, Nov.-Dec. 2000.

this last 0.5%, more than 98% is hydrogen and helium, while the remainder (a very small proportion) corresponds to all the other elements: those that make up most of the Earth and of our body. From this enumeration, some draw the conclusion that, since the atoms that make up our body are not abundant, we must be unimportant; our existence is an epiphenomenon; in other words, we are worthless.

The current attempt to humiliate man uses such stupid arguments, which actually work against the author's intention. Throughout history, scarce things have always been most valuable. When there are millions of copies of a postage stamp, it's worth nothing; if just three exist, they become invaluable. Velázquez and Van Gogh's paintings reach very high prices precisely because they are unique. Gold has the value it has, because it is one of the least abundant elements.

* * *

The biological dignity of the human species should not be asserted based on the number of genes (provisionally calculated at twice greater than that of the vinegar fly), or on their coincidences with other species, nor on the major or minor frequency of atoms, but according to their action on the environment, according to the phrase in the Gospel: *By their works you will know them*. Our dignity depends on our behavior, not on the number of our genes. Numbers are misleading. Can we claim that man is twenty times less important than elephants because man weighs twenty times less? Similar things are being heard lately, as a consequence of the tendency to quantify everything and confuse qualitative differences with those merely quantitative.

What are our works? What have we done to win the first position traditionally assigned to us, which only recently is being questioned? How do our works compare with those of other living beings?

Consider the Earth, seen from a distance. It looks stable, but it has changed considerably over time. Let's see how:

1. Before life appeared, the Earth was sterile. Its surface was divided, as now, in oceans and continents, but the latter were deserts. The dominant colors were brown and yellow. The atmosphere was mainly nitrogen and carbon dioxide. There was no ozonosphere, so ultraviolet rays reached the surface in much higher proportion than today. The apparition of life scarcely changed the aspect of the Earth, seen at a distance. The first living beings were microscopic, invisible in the range of scales we are considering.

2. About two billion years ago photosynthesis appeared, probably in beings similar to current cyanobacteria. Although they were still microscopic, these beings caused a visible change in the appearance of the Earth at a distance. First, in large masses they are green (the color of chlorophyll), which changed slightly the aspect of seawater. Secondly, the atmosphere became a mixture of nitrogen and oxygen, so its spectrum changed: an intelligent extraterrestrial would have noticed that the stripes in the spectrum corresponding to carbon dioxide had been replaced by those of oxygen. Third, some cyanobacteria grouped to live together in large numbers, forming macroscopic structures (*stromatolites*). For the first

time, living beings became visible to the naked eye, although there was no naked eye to see them.

3. With the passage from the third to the fourth level of life, i.e. with the appearance of plants (Metaphyta) and animals (Metazoans), the aspect of the mainland changed dramatically. The predominant color of continents became green. Animals are less visible from afar, but as soon as flying species began to emerge (insects, pterosaurs, birds, bats) they also become detectable. Finally, vertebrates changed the acoustic panorama of the Earth, filling it with sounds (trills, roars, etc.), which replaced the previous silence, broken only by the wind, the waves of the sea and the occasional landslide.

4. We come to modern man. Here we have an exceptional situation: for the first time in Earth's history, a single species can by itself change the aspect of the whole planet, seen from afar. The visual aspect of the Earth has changed completely, because the entire surface of continents is dotted with cities, roads, highways, ports, quarries... and what Selma Lagerlöf called *the big checked cloth*: the aspect from the sky of pieces of land dedicated to various crops. The changes we have produced are especially visible at night, when continents become sources of light through the illumination of cities and communication routes. The auditory landscape has also been modified, as man-generated noises are now dominant. But the most spectacular change is the fact that, from the 20th century, the Earth's radiofrequency spectrum has changed, becoming, by the action of man, a wave emitter in almost all frequencies. For aliens observing the Earth with radio telescopes, the effect must have been similar to an intense and sudden illumination.

This is not everything. Other groups of animals (reptiles, mammals) and plants have invaded all continents and can survive in all environments, including air and water, but different species occupy different ecological niches. Man is the only species that has done the same alone. About fifteen thousand years ago, a group of nomads crossed the Bering Strait, which separates Asia from America. The final colonization of the planet by man had to wait, however, until modern times. The Polynesians began their expansion across the Pacific Ocean towards the year 1000 BC. The arrival of the Maori in New Zealand did not take place until the year 900 of our era. Finally, the polar lands of Antarctica were conquered at the dawn of the 20th century.

Man is the only living being capable of consciously manipulating the evolution of other living beings. This has been done since thousands of years ago, through selective breeding, but lately the process has accelerated dramatically with the advances in genetic engineering. We will talk about this in chapter 12.

Likewise, man is the only animal species capable of causing a mass extinction, which threatens a considerable proportion of the other species of living beings. As we have seen in the previous chapter, there have been several extinctions in the history of life, but they appear to have been caused by meteorite impacts, volcanic eruptions, abrupt changes in weather, or other catastrophes. The extinction we are experiencing is different, because human beings have brought it about. For the first time, a single species causes, not just the extinction of its competitors, but of thousands, perhaps millions of species.

It may be argued that this is not something we should be proud of, that we should do everything to avoid it, that we are morally responsible to life on Earth for the damage we are causing. I accept it. But this also makes us different: we are the only living species that has raised this issue and has developed a moral obligation, which not only affects our relationships with other members of our own species, but also makes us accountable before all the others.

* * *

Man has done all these things, because he has transcended biological evolution and is the only species that has invaded a totally new field: cultural evolution. Culture has accompanied the human species throughout its history: *Homo habilis* was able to create primitive artifacts (the *pebble culture*). Progress was slow during two million years, but suddenly accelerated enormously.

The Neolithic revolution took place in the Middle East about ten thousand years ago: the inventions of agriculture and livestock made man independent of the environment and greatly increased the population. Cities, empires, civilizations appeared, then writing was invented. From that point, the genetic code ceased to be the most important means for the transmission of information between two consecutive generations. Biological evolution was replaced by cultural evolution.

The working of cultural evolution is quite similar to the working of biological evolution, but there are important differences. Natural selection, cultural variability, the influence of geographic isolation, act in the same way. Seen in this perspective, civilizations are equivalent to biological species. True, new phenomena arise: evolution gets faster, for it acts on the ductile cultural environment;

hybridization, a difficult phenomenon among different zoological species, is common among human cultures and civilizations. In the late 1960s and early 1970s, there were a few attempts[49,50 and 51] to study parallels between both forms of evolution: biological, shared by man with all living beings; and cultural, unique to us. Also important are the works of the Russian-American biologist Theodosius Dobzhansky (1900-1975), compiled and published in 1983 under the title *Human culture: a moment in evolution*[52].

I have participated in this work. In 1977 I wrote a book[53] where I proposed the theory that the elements of human cultures are subject to the laws of evolution, with a few differences. In essence, I distinguished two levels, one lower (biological) and one higher (cultural). The thesis of the book is summarized in its introduction as two fundamental principles:

1. The laws of evolution applicable at the lower level also apply at the higher level.

2. The laws of evolution applicable at the higher level don't necessarily apply at the lower level.

About this time, Richard Dawkins[54] invented the name *meme* to represent the ideas and fundamental elements of culture, which he

---

[49] Campbell D.T. 1965. *Variation and selective retention in sociocultural evolution*. In Barringer H.R., Blanksten G.I. and Mack R.W. (eds). *Social change in developing areas, a reinterpretation of evolutionary theory*. Schenkman Publishing Co.

[50] Cavalli-Sforza L. and Feldman M. 1973. *Cultural versus biological inheritance: phenotypic transmission from parents to children*. Human Genetics 25: 618-637

[51] Cloak FT. 1975. *Is a cultural ethology possible?* Human Ecology 3: 161-182.

[52] See note 46.

[53] *Human Cultures and Evolution*, Vantage Press, New York, 1979.

proposed as the cultural equivalents of *genes*, and founded the science of *memetics*, which studies evolutionary processes in the field of culture. Some biologists, such as Stephen Jay Gould, consider *memetics* just a metaphor with little value. I believe, with Dawkins, that it is something more, but it's true that the analogy can fail, as a consequence of the desire to prove that biological and cultural evolution are exactly the same, forgetting the differences. From the detailed study of cultural phenomena, it follows that many evolutionary mechanisms do act, really, in the same way on genes as on memes (this is what I meant with my first principle), but new, emerging phenomena and processes also appear, which differentiate both fields of application, according to my second principle. The problem with Dawkins's *memetics* is that it usually ignores emerging phenomena and processes.

One of the most significant differences, which affects cultural elements, but not biological features, is the concept of truth, which provides a criterion of natural selection unknown among genes. Indeed, it cannot be said that one gene is truer than another gene, it can only be more useful for the survival of the individuals who possess it. Among the memes, however, there are some for which the criterion of truth or falsehood becomes essential.

Consider the numerous examples that science provides: the reason why Newton's theory of universal gravitation was supplanted by Einstein's general relativity, after two centuries of uninterrupted triumphs, was because the second one comes closer to the truth (describes best the universe). Obviously, a truer theory also has, in some sense, a greater utility, but not necessarily in every

---

[54] Dawkins R. *The selfish gene*, Oxford University Press, 1976.

application. Established theories, although false, can provide greater political or economic advantages than those associated with correct scientific predictions. However, most scientists and philosophers of all time have argued that it is man's duty to find and defend truth against all benefits of any other kind that can be obtained from false or incomplete theories.

This is the fundamental failure of Dawkins's *memetics*, and also of Daniel Dennett's theories: they try to explain cultural evolution through the typical mechanisms of biological evolution, without taking into account emerging concepts such as duty or truth. Naturally, they argue that these concepts are also *memes*, therefore subject to *memetic* handling, without noticing that by saying this they incur in contradiction, for concepts such as natural selection or evolution, which support their theories, are also *memes*, and yet they are excluded from their analysis, being taken as proven, as axioms. We know that every scientific theory requires a certain number of fundamental axioms or postulates, which need not be demonstrated. I argue that the axioms of *memetics*, according to Dawkins, are incomplete, and that the emerging concepts I have mentioned are inexplicable in terms of pure biology.

Starting from their militant atheistic beliefs, Dawkins and others use *memetics* to attack religion, which they classify as a parasitic *meme*, similar to a harmful virus. It is surprising that scientists can forget, for ideological reasons, the fundamental criterion of science: truth, which must be set above utility. This is noticeable even in the titles of their articles, such as *What evolutionary good is God?* published by Dennett in the magazine *The Sciences* in January 1997. There is no question in the article of the truth or falsity of God's existence, just the usefulness of the concept.

In an article published in 1944, C.S. Lewis[55] addressed the loss of the sense of the truth and falsity of arguments and discussions, which I consider one of the most dangerous threats for future scientific development. A typical way to oppose a religious or philosophical idea consists in assuming, as a starting point, without proof, that the idea is absurd or false. From that point, one tries to describe the non-rational causes that brought the other person in the debate to acquire that belief or opinion. In this way, attention is distracted from the fundamental question, the only one that should matter: is the idea really absurd or false?

This is exactly the method used by Dawkins and his followers to attack religion. In fact, it is a version of a classical fallacy, well-known since the Greeks, whose name is the *fallacy ad hominem*. It is used to prove anything by twisting and abusing the intellectual human faculties that made possible the true advances of thought and science.

I can understand that agnostics may not be sure of the existence of God and want to keep an open mind. Believers have no problem, since they rarely support their faith in God's existence on scientific reasons. But I find it difficult to understand the position of militant atheists, who cannot prove the inexistence of God, but always take this assertion as their axiom or starting point, pretending that it is a result proven by science. It is easy to understand, by thinking just a little, that science cannot prove that God does not exist. Pseudoscientific atheism is just another form of religion, disguising itself as a science[56]. We'll get back to this in chapter 11.

---

[55] C.S.Lewis, *Bulverism*, The Socratic Digest, pp. 16-20, June 1944. In *God in the dock*, Eerdmans, 1970.
[56] There are other atheistic religions, such as *Hinayana Buddhism*.

* * *

In summary: from all the studies mentioned, it follows that the evolution of life has not changed fundamentally: the same current underlies all changes in beings of the first level (chemical evolution), second and third (cellular evolution), fourth (organic evolution) and man, considered as the creator of culture (cultural evolution), although by jumping to a different level, new phenomena arise, unpredictable at the lower levels.

It is also evident that, with the apparition of man, a threshold has been crossed, a change of state has taken place. In the same way that water, when crossing its critical melting point, passes from total immobility (or extremely slow mobility) to the freedom to move in two dimensions; in the same way that water, when crossing its critical boiling point, passes from mobility in two dimensions to the freedom to move in the three dimensions of space; in the same way, life, by crossing the critical point of reason at the apparition of man, passed, from a purely genetic evolution, to the freedom, inconceivable for all previous species, to evolve in a new field, the cultural environment. This is a much faster and more flexible form of evolution, which allows exchanges between different species (civilizations) and the adaptation of a specific human being to a totally different culture, which would be equivalent, in principle, to the impossible phenomenon of a living being changing, during its life, the species to which it belongs.

The unique character of man, as a species creative of culture, is incontrovertible, to the point that a few modern biologists hold that man should be classified in his own kingdom (in parallel to animals and plants), as his ecological impact on Earth has been at least as large as that of plants, and larger than that of all animals together. I think there is overwhelming evidence that, as I said at the beginning, *man is the current summit of evolution*. Of course, this does not mean that it must continue to be so in the future, that *we have reached the end of life's history*.

7. Towards the fifth level

Is there a fifth level of life? Haven't we reached, as sketched in the previous chapter, *the summit of evolution*, namely *man*? Is there now something even more complex, more important, on Earth? Can there be in the future?

We have around us unmistakable indications that evolution is ready to take another leap. This is reasonable. Why should the fourth level be the end of the process? Why shouldn't we have a new leap, after having had one from the first level (nucleic acids) to the second (prokaryotic cells); another from the second to the third (eukaryotic cells); another from the third to the fourth (multicellular beings)? If a change in level has happened three times, why not four?

Summarizing what we have seen in the previous chapters, the leap from one level of evolution to the next always consists in the union of several individuals of the lower level to form a more complex super-organism. A prokaryotic cell contains many molecules of nucleic acids; one eukaryotic cell contains several prokaryotes; one animal or a plant is made by the union of a multitude of cells. It is evident, therefore, that a hypothetical being of the fifth level should be made by the union of many individuals of the fourth level (plants or animals) acting jointly to assure the common good of the group they all make.

I have said at the beginning that we have indications that the appearance of beings of the fifth level is not a mere philosophical lucubration. There are facts that point unmistakably in that direction. Which are these facts? What incipient individuals of the fifth level exist today on Earth?

In the warm seas of the globe, especially in the western Pacific, there are structures, once the terror of the sailors, called coral reefs. They consist predominantly of calcium carbonate (limestone) that comes from the skeletons of countless tiny beings called polyps. At moderate depths, not more than one hundred meters, the polyps are alive and form colonies. All the individuals in a colony are closely related, for they come from a single ancestor that has reproduced by budding. This method of asexual reproduction works in the following way: the body of the parent produces buds, which develop and become complete individuals. Budding is typical of relatively simple aquatic invertebrates, such as Radiata, Porifera, Bryozoans, Tunicate chordates (ascidians), and lesser known groups. Sometimes, the individual formed from the bud separates and makes independent life, but in other cases remains physically attached to its parent, and the repetition of the process leads to the formation of a colony.

The colony functions as a unique super-organism: component individuals are linked together through their own body, have common parts and, in some cases, specialize in the performance of specific functions. There is a group of Radiata, the Siphonophorae, whose colonies float in the sea and whose appearance resembles jellyfish. The individuals forming the colony are divided into several types that differ a lot from each other in form and function. Some are responsible for ensuring the buoyancy of the colony and have become bladders full of air. Others lengthen and become tentacles or filaments loaded with stinging cells, which are used to capture prey, which is digested by other members of the colony. A few, finally, take over the reproductive functions.

Colonial invertebrates are very old: their first remains date back from the Cambrian period, more than five hundred million years ago. Does this mean that there were beings of the fifth level since the beginning of the Phanerozoic?

I think the answer must be negative. Colonial invertebrates are elementary organisms, quite close to the beginning of metazoan diversification. The individuality of multicellular animals is not quite developed at that level of evolution, so the boundaries between individuals are fuzzy. The members of a colony act more as organs of a complex metazoan than as building blocks for a higher order entity. In any case, we must recognize that, although the results are not spectacular, this is an incipient trial of evolution towards the fifth level of life.

Social insects offer the second example, much more to the purpose. Two orders of these animals evolved during the Mesozoic era to give rise to four different groups of species that live in society. The process culminated, apparently, during the Cretaceous period, and in the last hundred million years there have been no appreciable changes in those four groups. The two orders mentioned are Isopterans (termite) and Hymenoptera (ants, wasps and bees). The evolutionary ancestors of these two orders separated long before the onset of social trends. This is, therefore, a case of adaptive convergence, a typical process of biological evolution, as a result of which the same results can be achieved independently by several different paths. A classic example of adaptive convergence is the mole (a mammal) and the mole cricket (an insect), whose physical form is very similar, as both have independently adapted to underground life.

Isopterans are related to primitive cockroaches of the Upper Carboniferous period, and reached the social stage at least in the Triassic, two hundred million years ago. Hymenoptera had a later origin and are the most evolved insects. They seem to have reached social life several times independently, for in this group there are social, and also non-social species (bees and wasps that make a solitary life).

In the societies created by these animals, we can see the same phenomena of differentiation, dependence and solidarity that we have observed always come with changes in level during the evolution of life:

- Differentiation: individuals of the lower order specialize to perform different functions, required for the survival of the higher level organism.

- Dependence: the union becomes so close, that each individual cannot survive if, by chance, it is separated from the whole.

- Solidarity: the primary objective of every member is to ensure the proper functioning of the whole.

Insect societies get functional differentiation in several ways, depending on the group. Termites, for example, may join above one million individuals belonging to various castes: fertile males and females (kings and queens), provided with wings, responsible for reproduction; sexually inactive males and females, which in turn are divided into soldiers, who deal with the defense of the colony, and workers, who perform various functions related to feeding and building the termite mound. In some species, both can be subdivided into other sub-castes. There are also fertile males and females, substitutes for the king and queen, who can belong to several intermediate castes between them and the workers.

Termites are not born belonging to a particular caste. Their differentiation takes place as a result of the food they receive in the larval stage. In particular, the king and queen secrete hormones that inhibit the production of other fertile individuals. If one of them dies, that hormone stops spreading through the colony and some of the younger larvae develop into a new king or a new queen.

Social hymenoptera are bees, wasps and ants. They are the most intelligent insects (if one can speak of intelligence at that level), and they have a well-developed brain, greater in proportion than the brain of other insects.

Ants, closely related to wasps, are all social. An anthill can contain, from a few dozen individuals, to more than half a million. The number of castes also varies, depending on the species, from three (fertile males and females, and workers or sterile females) to more than twenty. As in termites, the food that a larva receives determines the caste to which it will belong.

Wasps and bees include social and solitary species. Social wasps form nests of paper or mud that can contain up to ten thousand individuals, while beehives house up to fifty thousand. The number of castes of these hymenoptera is just three: fertile males and females, and sterile females (workers). Social differentiation is also, in this case, of trophic origin (it depends on the food each individual receives in the larval stage), but in the case of bees it does not translate in functional body differences, for their behavior depends on their age: the same worker first cleanses the hive, later feeds the larvae, then watches the entrance to the nest, and finally collects pollen and nectar from near flowers. These behavioral changes are not a consequence of education, but the result of genetic programming, therefore it's common to the species. Thanks to this, a single caste successively performs the tasks that, in other social insects, require the existence of different castes.

Among the ants, wasps and, especially, termites, there is a form of communal feeding resembling the common body of polyps, where each individual digests only a part of the food it eats and excretes the rest, which will be eaten by another individual. In this way, the entire colony is fed by just a few, giving rise to a collective stomach. In some cases, this results in the emergence of a curious forms of parasitism, as in Amazon ants (*Polyergus*), whose workers are specialized in fighting, and would starve, even in the presence of food, unless a worker of another species, *Formica fusca*, feeds them. The Amazon ants attack the nests of *Formica fusca*, kill their queen and enslave the workers. In even more extreme cases, as in ants in the genus *Anergates*, the queen invades a nest of the genus *Tetramorium*, supplants their queen and, fed by the workers of the other species, produces eggs that become queens and males of its own species, without ever generating workers, as they are not needed.

In bees, however, there is no collective stomach: honeycombs play that role. The nectar collected by the workers, transformed into honey by the action of digestive enzymes, is regurgitated and stored in the hive. When a bee is hungry, it can be satiated in anyone of the honeycombs.

A termite mound, an anthill, a wasp nest or a beehive, is the closest on Earth to a fifth level individual. For all practical purposes, these societies function as a single body. Each of its members can only live in contact with the others. If an ant is lost, it dies in a very short time. If it finds another anthill, even of the same species, it cannot become a member, as its different smell will prevent its being accepted.

The future of evolution is not likely to pass through social insects. In fact, there are indications that their evolutionary development has stopped. Ants and fossil bees are known, practically identical to the present species, having lived more than thirty million years ago, which probably behaved as they do now. It's not difficult to discover the reason. Insects are restricted, due to their physical structure, to very small sizes. The presence of an external skeleton (a chitin shell) is an important impediment to their growth. An internal skeleton is much more efficient, therefore vertebrates are the only land animals that reach large sizes. Science fiction movies based on invasions by huge ants or spiders, are not possible in practice: the exoskeleton of the giant ants would be unable to support their weight, and their legs would break.

Being small may have advantages, but it also has disadvantages: the brain of a tiny animal cannot have the same number of nerve cells as that of a large one. Its intelligence, therefore, is limited. Their behavior must be, for the most part, instinctive; genetically determined. Evolution won't have space to act. Sooner or later, a dead end is reached.

This is what has happened with social insects. With them, perhaps evolution has reached the highest levels of instinctive complexity that can be reached by a limited nervous system, as arthropods have. As a proof of this, in so many million years since the origin of these societies, biological evolution has just produced secondary changes, both in isopterans and in hymenoptera. These changes have given rise to a great diversity: two thousand species of termites, more than three thousand of ants, more than one thousand of wasps and social bees; but there seems to have been no appreciable progress in their social structure. They are successful animals, spread throughout the world, but stagnant. We won't find among the insects the pointer showing us the path towards the future of evolution.

* * *

The tendency towards the formation of societies is also evident among vertebrates: banks of fish; flocks of birds, with a more or less strict hierarchy; the complex structure of packs and herds of mammals (deer, antelope, bison, wolves, baboons and many more); everywhere we find a repetition of the same phenomena: the appearance of an organized group, functional specialization, which in vertebrates is normally limited to their behavior, without affecting the body structure of individuals, except in regard to sexual differentiation. It is true that these societies cannot be compared in complication and efficacy with arthropod societies. On the other hand, they are much less rigid: there is a lot of vertical and horizontal mobility in the herd, flock or pack. Any well-endowed individual can, in principle, climb to the highest levels of the hierarchy of the group by their own efforts, unlike what happens in social insects, where the reproductive individual, at the center of the hive or anthill, is chosen in an apparently capricious way, once and for all.

But there is a vertebrate that has reached levels of socialization that don't lag behind those achieved by isopterans and hymenoptera. This is man.

During most of their evolutionary development, human societies did not differ significantly from herds and packs of mammals, either in the small number of members of the primitive tribe, or in its hierarchical organization.

The existence of the group was ensured through hunting and collecting spontaneous products of the earth. The first was risky, difficult and uncertain; the second also had dangers (attacks by wild beasts or neighboring tribes), and depended, to a considerable degree, on chance. Under these conditions, feeding demanded almost exclusive dedication of all the active individuals in the group. Only the elderly, the sick or disabled, and the children, would enjoy some free time. The sick were scarce, as there was a very high mortality rate and their half-life didn't go beyond thirty years. So it is not surprising that the few who managed to exceed this age, and therefore had a considerable experience, compared to their younger colleagues, were mainly dedicated to government and the administration of justice. A tribal society was therefore made of three or four categories of people, who differed by their work: hunters (usually men), gatherers (often women), elderly people, and otherwise useless people, which could be devoted to specialized tasks (carving stone, healing the sick, etc.). Belonging to one or another class was determined by the age, sex and physical built of each individual.

About ten thousand years ago, with the Neolithic revolution, cities and modern states emerged, societies much larger and more complex than the tribe, integrating thousands, millions and even hundreds of millions. In extreme cases, many more than the largest groups of ants or termites.

At the same time that this happened, social differentiation grew. Feeding the population depended now on agriculture and livestock, less dangerous occupations than hunting and gathering, with more predictable results. On the other hand, the new techniques of farming and animal husbandry made it possible for a single man to produce enough to ensure the support of many. It was not necessary that everyone should be engaged in tasks of this kind. Free time abounded and it must be used somehow.

One of the first specializations was the warrior, who provided immediate benefits to society as a whole: the army protected farmers and cattle breeders against attacks from rival neighboring societies, thus reducing the risks of the trade and increasing the efficiency of other activities. In exchange for their protection, the workers delivered a part of their products (taxes) to maintain the warriors who defended them.

Like every human and animal group, primitive armies acquired a hierarchy. The supreme commanders found themselves in a privileged position: their men were faithful and hardened, so they could be used for personal gain. As power encourages power, these chiefs progressively amassed most of the government of the city or territory whose subsistence depended on their protection. Thus arose the first military dictatorships. The innate tendency in the human being to prefer his own children had the inevitable consequence that the privileges of the supreme positions became hereditary. Monarchies and empires appeared.

Society was divided into castes or social classes: the children of the soldiers became soldiers, those of the peasants cultivated the land or raised domestic animals. In some places (notoriously in India, after the invasion of the Aryan peoples) custom became law and vertical mobility decreased almost to zero. In other countries, a certain degree of social elasticity was allowed: a determined man had some hope of rising through his own effort.

Parallel to the warrior caste, a religious elite developed. The more or less animistic tendencies of primitive man, which seem to go back even to our relatives, the Neanderthal, underwent a transformation with the Neolithic revolution. The success of crops is uncertain, depends on the climate, and concern the survival of the society, while the protection of the army is useless against this danger. Some things, as the seasons, are predictable, for they are linked to the position of the sun in relation to the stars. Therefore, the first civilized men reasoned, the apparently unpredictable atmospheric phenomena, storms, hailstorms, droughts and catastrophic floods, must also be related in some way to the movements of the stars, or else they will be a consequence of the whim of a god. Both answers were applied. The second (probably older) led to the appearance of a priestly caste charged with appeasing immortal beings, on whose goodwill the subsistence of the nation would depend. The first, closely related to the other one in the places where it flourished (Mesopotamia, Central America and China) led to the development of divination techniques of various types and to the first science of modern man: astrology.

The priestly and warrior castes had some things in common. Both were exempt from productive work, so as to be able to devote themselves exclusively to the protection of the producers against their various divine and human adversaries. But the priests, as a consequence of their activities, were the most educated among the inhabitants of the first civilizations. It is not surprising that they were requested to keep accounts, not only of celestial movements (the calendar), on which the crops directly depended, but also of the collection of taxes and their distribution among the non-productive classes. This priestly activity led to the invention of writing systems and numerical notations, as well as the development of arithmetic, the second science of modern man. This happened independently in Mesopotamia and Egypt (about 5000 years ago), India (4500 years ago), China (3500 years ago) and Central America (more than 2500 years ago). Shortly after, a new caste or profession emerged in most of these civilizations: the scribes.

A part of the population, whose efforts were no longer necessary for the production of food, devoted themselves to technological activities and gave rise to new specializations: guilds or castes of artisans. In this group, the oldest professions were pottery and metallurgy: first copper, then bronze, later iron. At first, the members of these professions used to be people physically incapacitated for another type of activity, as also happened in tribal societies. This has left its mark in old mythologies: often, the god of blacksmiths is lame (Hephaestus in Greece, Vulcan in Rome).

The proliferation of products, manufactured or cultivated by specialists, caused the appearance of traders. First by bartering, later by means of metallic currency, invented in Lydia around 700 BC, the members of this profession favored the dissemination of the fruits of human labor, sometimes exceeding the limits of civilizations.

Social differentiation did not stop there. The number of castes, professions, social classes... multiplied. Hinduism, the dominant religion in India, classifies human beings in four hundred thousand different castes (there is one even for thieves), each of which has its own moral code, and strict laws that regulate mixed marriages. Without reaching such extremes, the number of different professions (socio-cultural specializations) that can be found in a modern civilized society is enormous. Sometimes, their membership has become hereditary. The Roman emperor Diocletian (245-313) enacted a law that forced boys to embrace the same profession as their father. Even today, strong social or family pressures try to get the same result, even if it's not compulsory by law.

At present, at least in countries belonging to the Western civilization, the freedom of each individual to choose a profession is generally accepted. Social specialization is no longer hereditary and has become cultural. But, as we saw at the end of the previous chapter, the laws of evolution act similarly on the two substrates, although there also are some important differences.

***

As members of human society differ from one another (not physically, but culturally and professionally), their dependence on the social group increases. Before the Neolithic revolution, a tribe of primitive men was self-sufficient. It probably happened frequently that a single couple became independent and formed a new tribe of their own in a previously unoccupied territory. Two was the minimum number of human beings capable of living alone and reproducing to found a new society.

A modern society, on the other hand, educates its members in such a way that they are helpless and perish if they are forced to live in isolation. In the best case, if they survive, they'll be reduced to savagery. Although based on a real event, Robinson Crusoe[57], who manages to live completely alone for more than twenty years on a desert island, is less realistic than Ayrton[58], who in the same circumstances quickly loses his humanity, becoming practically a wild animal.

In the eighteenth and nineteenth centuries, the euphoria produced by the avalanche of scientific discoveries that gave rise to the *myth of indefinite progress* led to the romantic idea that civilized man would be able to dominate nature, whatever the circumstances. We have seen that Verne does not fall into this fallacy in the case of a single man, but even he gets carried away when a few men are together. Precisely in the same novel, *The Mysterious Island*, where Ayrton brutalizes in total solitude, we also find engineer Cyrus Smith who, with just four companions, successfully solves the same challenge.

---

[57] Daniel Defoe, *The life and strange surprizing adventures of Robinson Crusoe, of York, Mariner. Written by himself*, 1719.
[58] Jules Verne, *Les enfants du capitain Grant*, 1867-68, *L'île mystérieuse*, 1874.

What Verne wants to say with these two opposite answers is obvious: where a single man would be doomed to savagery, a small group could avoid such a fate, especially if a *deus ex machina* is helping them, a role played by Captain Nemo in *The Mysterious Island*.

A special case concerns children separated from adults and forced to survive outside human society, either by themselves or together with animals. We also find here the two possible situations: absolute solitude and a small group. The first is quite old. It is found, for instance, in the legend of the founding of Rome by two twins, Romulus and Remus, who having been abandoned as newborns, are breastfed by a she-wolf and fed by a woodpecker, until they are adopted by Pastor Faustulus. Romulus and Remus reach adulthood without difficulty, and the first becomes the founder of the city that will turn into an empire.

Two similar cases are given, in literature, by Mowgli[59], raised by a pack of wolves, a bear and a panther, and by Tarzan of the Apes[60], who loses his parents when he is one year of age and is adopted by a she-ape. Like Romulus, Mowgli and Tarzan become intelligent adults and end up fully integrated in society.

Reality seems to deny these optimistic visions. Historical cases of feral children, raised by wild animals, or those that have been deprived of intellectual stimuli during their childhood, as the mysterious Kaspar Hauser, indicate that human beings need to live their first years in certain conditions, among people who care for

---

[59] Rudyard Kipling, *The jungle book*, 1894-1895.
[60] Edgar Rice Burroughs, *Tarzan of the apes*, 1914.

them properly, for otherwise they won't reach normal intellectual development or adapt successfully to life in society.

As for the situation of a group of isolated children, we can turn again to literature, which offers Verne's optimistic version in *Two years' vacation*[61], and William Golding's pessimist one in *Lord of the Flies*[62], his best known work, that won him the Nobel Prize. Today we tend to think the second more realistic than the first, as the testimony of a few specific cases seems to confirm, although there is always some doubt. For obvious ethical reasons, we cannot experiment.

Each of us can imagine how we would manage to survive, if we were alone or part of a small group, on a desert island. If we are honest, we must accept that it would be quite difficult. Perhaps it could be said that the interrelationships between different human groups are so great now, that the minimum number of individuals capable of maintaining the current structure of society is practically the same as the population of the planet.

* * *

Summarizing: from the study of the history of humanity, it is evident that human society has been increasingly imposing on its members the (cultural) differentiation and dependence that always accompany a change in level in the scale of the evolution of life. Does the same thing happen with the third characteristic, solidarity?

Unfortunately, the answer cannot be totally affirmative. The cells of society, individual human beings, act freely (within certain

---

[61] Jules Verne, *Deux ans de vacances*, 1888.
[62] William Golding, *Lord of the flies*, 1954.

limits) and can oppose the good of the whole to seek their own selfish benefit, which social insects and polyps cannot do. Precisely for this reason, all human societies have complex legal systems that try to ensure that the uncontrolled exercise of individual freedom does not endanger the subsistence of the social group. This is the reason why every society is unfair to some extent, and favor some of its members to the detriment of others.

The lack of this fundamental requirement makes it impossible to consider the social group as a super-organism endowed with individuality. So we must answer in the negative to the question of whether human society, as we know it today, belongs to the fifth level of life. In an article published during the nineties, Heylighen and Campbell[63] offer a pessimistic point of view regarding the possibility that we will make the leap to the fifth level or, as they call it, a *metasystem transition*.

According to these authors, when there is a change in level, the evolution in the lower and higher levels move in contradictory directions. For each individual of the lower level, considered in isolation, natural selection should favor a selfish behavior, since their personal objective is the propagation of their genes to the next generations, which conflicts with the achievement of the same objective by different individuals. However, considered as a member of the higher level, this same individual should behave altruistically, renouncing its own benefit in favor of the higher order entity of which it has become a part. As we'll see in more detail in Chapter 11, some researchers think that the only possible

---

[63] Francis Heylighen y Donald T. Campbell, *Selection of organization at the social level: obstacles and facilitators of metasystem transitions*. In *World futures: the Journal of General Evolution*, 45, p. 181-212, 1995.

solution to the dilemma would be that the individuals of the lower level renounce their ability to reproduce, which would be reserved for just a few. In this way natural selection would stop acting at that level, and altruistic behavior could be favored, which would be best for the higher level, thus evading the indicated obstacles. However, both in the human species and in all social forms of vertebrates, there has been no such renunciation of reproduction by the individuals of the lower order.

Does it follow that our access to the fifth level is impossible? Not necessarily. Despite these consider-ations, it seems clear that human evolution is well advan-ced on the path that leads to the fifth level. Can we reach the end of the process? Will evolution use other methods to secure it? Is it desirable? What must we give up to get there? Can we foresee the features of a fifth level being, or at least those that won't be there, when evolution finally reaches that goal, if ever? The remaining chapters of this book will try to answer these questions.

8. The fifth level in literature

In the first chapters of this book we have summarized the current knowledge about the successive steps of the evolution of the universe. Starting from its origin, we reviewed the apparition of life on the third planet of the solar system and followed its evolution for more than three billion years until today, when the field of action of evolution has ceased to be predominantly biological and has become cultural.

In this chapter we are leaving the field of science (we will return later), and enter into more or less plausible speculations. The prediction of the future is always a risky attempt, because reality can show us spectacularly wrong, as polls usually indicate. However, making predictions is fashionable. Also the risk is less, the farther the predicted future; there is a certain probability that, by then, our wrong predictions will have been forgotten, precisely because they are far, while our right predictions are all the more impressive the farther away. That's why Jules Verne's scientific and social anticipations are so surprising[64].

On the other hand, we must take into account that most of the knowledge man has acquired has no other purpose than making possible to anticipate changes in our environmental conditions and estimate the consequences of human acts; in other words, predicting the future. It has been said that we are the only animals

---

[64] Jules Verne, *Paris au XXe siècle*, published in 1994, written about 1863.

capable of predicting our own death, which means that predicting the future is one of the essential features differentiating us from all other living beings.

If human society is a glimpse of the fifth level of life, as we have suggested in the previous chapter, perhaps we can risk asserting, without too much fear of being mistaken, that evolution will take us closer to that level in the future. Can we describe the way of life of the members of the fifth level? And if we cannot define it in detail, can we at least know how it isn't likely to be?

As usual, we'll start by asking the help of literature. It happens that the existence of the fifth level of life is not something that I, the author of this book, have invented. It has been discussed for thousands of years in books belonging to an easily recognizable literary genre.

In fiction literature, and in futuristic sociological essays, many authors have attempted to describe perfect societies where all human beings cooperate harmoniously for the common good. The literary genre dedicated to the description of these societies is named after one of its most famous examples, *Utopia*[65], by the British writer Sir Thomas More (1478-1535). After what was been said, it's evident, that utopias are nothing but the vision that the corresponding author had about the fifth level of life. That vision varies considerably from one author to another, but there are many common elements that can provide us with useful ideas about the question we are considering.

---

[65] From the Greek *u*, negative particle, *topos*, place. *Utopia* means thus *what is in no place*.

There is another group of literary works, more or less contrary to utopias, which we can call *dystopia*[66], name applied to those works that present exaggeratedly imperfect societies, with the intention of criticizing the vices of the society where the author lives, or warning human beings about undesirable tendencies in the path of evolution.

The first proper utopia was Plato's *Republic*. In this dialogue, Plato (c. 427-c. 347 B.C.) makes Socrates formulate the famous statement that *the rulers of the perfect state must be philosophers*. The society itself would be divided into classes: the *people*, who perform the necessary work for survival; *soldiers*, who defend the society against attacks by foreign societies; finally, leaders, whom Plato calls *guardians*.

The Platonic Republic will stay united if each social class behaves according to its proper virtue: the people must practice temperance; the warriors, fortitude; the guardians, prudence. The whole is harmonized by the collaboration of citizens in the supreme virtue: justice. The members of the republic will share all their properties with each other (for Plato, property included women and children).

The perfect republic would start to exist almost automatically, according to Plato, as soon as a philosopher gained power or a leader became philosopher. In his own words: *If a ruler introduces [ideal] laws and social conventions, it wouldn't be too much to expect that his subjects will consent to submit to them*. Once this type of society is instituted, it will be maintained indefinitely, throughout generations, by means of a system of education (each

---

[66] From the Greek *dys*, bad.

subject according to his class), combined with a rigid censorship that eliminates all possible deviations and prevents dangerous thinking. The artistic temperament, too difficult to reduce to norms, is one of the activities that the Platonic Republic would forbid, to ensure its own stability.

The great Greek philosopher not only drew the master lines of the perfect society: he tried to put them into practice. He traveled to Syracuse at least twice, and tried to educate the son of the tyrant Dionysius the Elder (c. 430-367 BC) and make him a philosopher king. The attempt failed. Some say that Plato ended up being sold as a slave, and that one of his friends, Anniceris of Cyrene, had to ransom him out, but this has not been confirmed by historical research. As for his disciple, Dionysius the Younger (396-c. 330 BC), he did not do much honor to Plato's teachings and governed Syracuse sporadically, for he was exiled twice.

It is curious that Plato's Republic, ruled by philosophers, was actually put into practice, centuries after his death, by the Roman dynasty of the Antonine. In particular, Emperor Marcus Aurelius (121-180) was educated in the teachings of Stoicism and became one of the main representatives of this philosophical school, founded by Zeno of Citium (c. 336-264 BC) and Chrysippus (280-207 BC), to which also belonged Epictetus (55-135) and Lucius Annaeus Seneca (c. 4-65).

Marcus Aurelius tried to apply the Stoic teachings in his government, by attempting to act always in accordance with the dictates of reason and being indifferent to passions. In general, his decisions were beneficial. He lowered taxes on the poor, interpreted benignly the Roman law and offered protection to slaves, orphans and poor people. On the other hand, he persecuted Christians. Some consider this a deviation from his usual behavior; however, his attitude is consistent with Plato's theories about the perfect republic. As noted by Toynbee (1889-1975) in his monumental *Study of History*[67], what he did was applying state censorship to the control of dangerous thinking.

Where Marcus Aurelius did fail was in the last and most momentous decision of his life. Unlike his predecessors in the dynasty, who ignored blood ties and used the Roman institution of adult adoption to choose as successor the most suitable person, Marcus Aurelius appointed as heir his own son, Lucius Aelius Aurelius Commodus (161-192), clearly useless for the royal office, who would rather act as a gladiator than as a ruler. He was strangled in his bath by the athlete Narcissus. With him the dynasty of the philosopher kings ended, and Rome sank into an anarchy that lasted about a century.

The fundamental defect of the Platonic Republic is its instability. It makes a precarious balance that, ultimately, can only be maintained by force, censorship and the elimination of personal initiative. In this utopia, as in most of them, individuals must submit totally to society, whose ultimate goal becomes its own survival. To ensure this, the objective of making its members as happy as possible becomes secondary. But then, human beings won't consent to take the necessary steps to build such a society, or keep it running, if it already exists.

* * *

---

[67] Arnold Joseph Toynbee, *A Study of History*, 1934-1954, in twelve volumes.

We now turn to the paradigm of the works in this genre, *Utopia*, written in Latin by Thomas More on the occasion of his diplomatic trip to Flanders in 1515. The work was published the following year and is a fictional travel story, where a sailor, a Portuguese companion of Amerigo Vespucci, describes the countries he has visited, especially the perfect society of Utopia, supposedly located in the new world and in the southern hemisphere.

According to More, the perfection of the society of Utopia is based on the total absence of private property and money. Everything is owned by the community, and citizens share things according to their needs. Dwellings, for instance, are awarded by lottery and changed every ten years; in each home lives a family of ten to sixteen adults and an undetermined number of children, under a patriarchal regime. The remaining goods, food, clothing and work tools can be taken from community stores just by asking for them. It is assumed that no one will try to accumulate more than they need.

Work in Utopia is not exhausting: six hours a day for every man and woman. The rest of the time, they are free to devote themselves to educational activities or healthy recreation, always monitored by the person responsible for their family, to ensure that no one remains idle or wastes time in non-recommended occupations. Citizens must also work for two years in social service by carrying out agricultural tasks, and must submit to military training during the two feasts they have every month.

Although More denies the existence of social classes in Utopia, things are not so simple: there is a mass of people engaged in manual labor (masonry, carpentry, blacksmithing, etc.), but the most unpleasant tasks (cleaning, garbage collection, slaughter of animals) are reserved for slaves, citizens of Utopia who have committed a crime, or foreigners hired for this purpose.

On the other hand, there is a third *de facto* social class in Utopia: *intelligentsia*. In each city of six thousand families, some five hundred people are excluded from manual labor, because their exceptional intelligence makes them fit to engage in scholarly trades. This group includes diplomats, priests and members of the government, elected by secret ballot of citizens, who are entitled to certain privileges, such as better food; but anyone who deliberately tries to get into a public office is permanently disqualified for this position.

The control of the activities of the population rests with two hundred district officials, each of whom supervises thirty families. They must make sure that everyone works the proper hours. Solving public litigation and other government issues corresponds to a committee of twenty members (*senior district controllers*), chaired by the mayor of the city, who is elected by the assembly of district controllers. This position is for life, while other administrative positions are renewed every year, although in some cases re-election is possible. In addition, there is a general parliament for the whole country, with three representatives per city, which deals with national issues and keeps the balance of the population and resources for the fifty-four cities of Utopia, ordering transfers, if required.

The stability of the Utopian society is ensured by the following control elements:

1. The belief of the citizens in the immortality of the soul, and their reward or punishment in the other world, in accordance with their works. This makes them fulfill their duty towards society, even when such duty is unpleasant or dangerous, since the structure and laws of the community get religious sanction. In a prayer, recited monthly in public by all citizens, they tell God that *if our government is the best and our religion the truest, [keep me faithful to both] and bring all the world to the same rules of life.*

2. Education: All children are raised in way that seeks to instill, from the earliest childhood, the main principles of their society. Priests, in particular, are responsible for the education of children, and do their best to ensure that *children [are infused] such opinions as are both good in themselves and will be useful for their country.* In other words, they are taught the ideas best-calculated to preserve the structure of their society.

3. Public opinion and constant control by administrators, family members and neighbors: *Everyone is watching you, so you are practically forced to do your work and make proper use of your free time.* Ridicule is also a weapon used to deter those who behave in an unorthodox way.

4. Various public honors to people who have distinguished themselves in the performance of their duty, or who have rendered extraordinary services to the community: For example, *they erect statues... in the marketplace... [as] incitement to their posterity to follow their example.*

5. Active deterrence by the immediate punishment of crime, even if it wasn't completed: Deliberately trying to commit a crime is legally equivalent to having committed it. Penalties range from

slavery (for serious transgressions) or mandatory celibacy (for premarital sex), and even the death penalty (for relapse or rebellion).

When judging More's *Utopia*, we must take into account the structure of the British society of his time, dominated by a selfish, hereditary ruling class, while most of the population was in almost absolute misery, without protection in the case of old age, physical disability or unemployment. Remember also that the legal system punished almost any crime with the death penalty, even the smallest, so that honest workers left without work became beggars, and then they were executed if their desperate situation made them steal to eat. In *Utopia* More was reacting against this state of affairs, presenting a more just social situation, compared to his own.

Anyway, More saw that human beings are individually unpredictable, selfish and inclined to evil. Therefore, he provided his society with incentives and deterrents, to ensure that most citizens will fulfill their duty, thus making possible its stability. Hence also the importance of education, and the fundamental role played by mutual control and vigilance among citizens, one of the most unpleasant characteristics of *Utopia*. In spite of everything, More was not so naive as to assume that those procedures would be enough to ensure the submission of all individuals to the society, so he foresees the constant presence of aberrant elements, who are punished by slavery or death, if needed.

Utopian society contains in itself the seeds of instability. In practice, such an organization could not remain in balance during the 1760 years that More says it has endured. In the first place, it isn't true that the abolition of money and private property would make human greed impossible. During the first phases of the evolution of humanity, before the invention of currency, when all trading was made by bartering, there also were accumulations of goods or wealth in a few hands. As soon as a scarce product would appear in *Utopia*, internal tensions would develop between the citizens eager to have it or use it. On the other hand, by giving privileges to a ruling class, More makes ambition possible. Although he says that the desire to occupy one of the key positions is enough to disqualify the candidate, it is obvious that a skilled man could get there, pretending he's doing it against his will, following the wishes of his fellow citizens. In eighteen centuries, there is more than enough time for these things to have happened, not one time, but many. Once an ambitious, determined man had reached an important position (such as lifelong mayor), the desire for power could lead him to impose a dictatorship, so that the social structure of *Utopia* would disappear forever.

* * *

The technique of using a traveler's story as a framework for the description of a society different from ours was not unprecedented when More made use of it, and has been exploited countless times after *Utopia*. But as the exploration of the planet reduced the extent of *terra incognita*, the authors encountered increasing difficulties to find a place for their utopias. In 1602, Tommaso Campanella (1568-1639) wrote his *The City of the Sun*[68], clearly influenced by More's *Utopia*, both in its literary form and in the type of society described: this is also the story of a sailor, in this case a companion

---

[68] *La città del sole.*

of Columbus, and is located on a large plain, just below the Equinox, not far from Taprobana (Sri-Lanka).

In 1872, Samuel Butler (1835-1902) placed his dystopia Erewhon[69] in an unidentified British colony, whose description reminds of New Zealand. The lost islands of the Pacific Ocean[70,71 and 72] or the lost valleys of the Himalayas[73] were also used in works of this genre. Finally, modern utopias and dystopias had to place their societies underground[74], in other planets or satellites[75], or in the future[76].

The most influential futurist utopia of recent centuries was developed by the German economist Karl Marx (1818-1883), who interprets history exclusively in terms of economic causes and motivations (*historical materialism*) and reduces it to a struggle for supremacy among social classes, which is reflected in a series of violent revolutionary movements, which evict the upper class from its position and replace it by a new social group that, in turn, does not take long to become reactionary and oppress the remaining classes. According to Marx, all societies went through the same stages: tribalism, slavery, feudalism, capitalism and socialism. The last displacement in our western society would have taken place towards the end of the 18th century, in the early stages of the

---

[69] An anagram of *nowhere*, a word that means the same as *Utopia*.
[70] *New Atlantis* by Francis Bacon (1561-1626), published in 1627.
[71] *Gulliver's travels*, 1726, by Jonathan Swift (1667-1745).
[72] *Island*, 1962, by Aldous Huxley (1894-1963), also belongs to this category.
[73] *Lost Horizon*, 1933, by James Hilton (1900-1954), describes the famous utopic valley of Shangri-La.
[74] *The coming race*, 1871, by E.G.E. Bulwer-Lytton (1803-1873).
[75] *The First Men in the Moon*, 1901, by H.G. Wells.
[76] *The Time Machine*, 1895, by H.G. Wells.

industrial revolution (the French Revolution and its aftermath), when the feudal aristocratic ruling class was replaced by the bourgeoisie. But this class established a capitalist system maintained by a new oppressed social stratum, the factory proletariat, which was assigned the new kind of jobs created by the industrial revolution.

The society formed in this way is unfair, since the fruits of human labor do not revert to their rightful owners, the workers, who receive a salary notoriously lower than the value of their effort, usually set at the minimum necessary to sustain their life and family. The remainder or surplus value, is the exclusive property of the bourgeois (the owner of the company) and accumulates in the form of capital[77].

Marx then predicts the future development of human society, which he considers evident and inevitable, because of historical necessity: class struggle, which in the eighteenth and nineteenth centuries confronted the bourgeoisie against the landed nobility, will happen again, confronting the capitalist bourgeoisie, triumphant in the previous confrontation, against the factory and rural proletariat[78]. The victory of the proletariat is certain, because they have on their side numerical advantage and unity (*workers of all countries, unite!*), by knocking down all barriers of race and nationality (*workers have no country*).

When the working class will reach their inevitable triumph, it will establish a despotic regime (*the dictatorship of the proletariat*), which will transform society in this way:

---

[77] *Das Kapital*, 1867-1894.
[78] *The Communist manifesto*, 1848.

1. Abolition of the private ownership of land, factories and means of production.

2. Abolition of the right of inheritance.

3. Centralization of the means of communication, transport, production and credit in the hands of the State.

4. Establishment of a free education system for all children in public schools. Abolition of child labor. Combination of education and industrial production.

5. Obligation to work, for all members of society.

6. Redistribution of the population, by elimination of the distinction between urban and rural areas. Combination of agriculture with industry. Establishment of agricultural shifts for all workers.

In the long run, these measures would destroy class distinctions and the current structure of society, which will be replaced by a classless society. When this occurs, all the institutions that served to ensure the dominance of one class over the others (such as the State and religion) will be unnecessary. The dictatorship of the proletariat will dissolve itself and become an association of free and happy human beings. The *utopia* will have been established forever.

Education will keep society together and ensure its stability. Education will be understood as a total activity, *extending over the whole active life*[79], rather than as a transitional period at the beginning of the life of a person. It will encourage individuals to participate in the common task, making them identify with the

---

[79] Roger Garaudy, *L'alternative*, 1972.

spirit of the new era, follow the path of evolution and become *a creative source of the future.*

What happens to individuals who do not share these ideals, who refuse to integrate into the new society? Marxism considers them as sick beings, who must be treated like any other mental patient; for no normal person, subject to proper education, can reject the Marxist utopia; or they may be reactionaries who wish to maintain their privileges, and should be defeated or, if necessary, eliminated for the good of the cause. To do this (at least in the early stages of the dictatorship of the proletariat), all the members of society must be constantly watched, to discover deviants and control them as soon as possible. Hence *district committees, commissars,* and similar institutions. The resemblance to More's *Utopia* is amazing, except for the theoretical absence of a ruling class.

We owe the best literary version of the Marxist utopia[80] to the British writer William Morris (1834-1896). Presented in the form of a dream, in which the author moves to the future (the beginning of the 21st century), he makes an idyllic description of the classless society, according to Marx's forecasts.

After Marx, Marxist thought has tended to become dogmatic and petrified. Marx's criticism against *communism and critical-utopian socialism* (his names) could be applied to his own followers with his own words: *Although the originators of these systems were, in many respects, revolutionary, their disciples... hold fast by the original views of their masters, in opposition to the progressive historical development of the proletariat* (Communist Manifesto).

---

[80] *News from Nowhere*, 1890.

The Fifth Level of Evolution

The course of history, in the century after the death of Marx, has not confirmed his predictions. The Russian Revolution, which many identified with the beginning of the dictatorship of the proletariat, failed after seventy years, but from the beginning it became, like all previous revolutions, a simple replacement of the old ruling class by a new one. Economic equality was not achieved. *In 1918, the Soviet government had to introduce different salaries in a proportion of 175 to 100... Since that date, inequality has grown relentlessly*[81].

In fact, the evolution of the social situation in capitalist countries has moved in part in the direction foreseen by Marx. The fourth point indicated above is today a reality. But it wasn't imposed by the dictatorship of the proletariat, although the action of labor unions has influenced to raise the proletarian class above the minimum level indispensable for life, where according to Marx, capitalism insisted on keeping it.

Human society, throughout its history, seems to show a continuous and invincible tendency to form hierarchies. As Russian-American sociologist Pitirim A. Sorokin (1889-1968) says: *The organization of a (social) group means... the division... into classes... and the emergence of some form of government. It implies, in turn, intergroup differentiation and stratification.* He gives four reasons for this:

a) The heterogeneity of individuals: we are not all equal, either biologically or socially.

b) The advantages of differentiation.

---

[81] Pitirim A. Sorokin, *Society, Culture and Personality*, 1962.

c) The facilitation of the social order: without hierarchy, society can fall into permanent struggle, where everyone rules and nobody obeys.

d) The incessant change of environmental conditions, which spontaneously raises some individuals higher than others. Karl Marx's perfect classless society would be unstable and evolve spontaneously towards a hierarchical situation, where the division into social classes would be reproduced.

The presence of a hierarchical organization is not exclusive of human societies: it is shared by all groups of higher vertebrates (birds and mammals), which means that the hierarchical group has been positively selected by evolution. It does not seem that the conditions have changed so much that the weight of this character in natural selection has been reversed.

Dialectical materialism, the fundamental basis of Marxist philosophy, must be considered with caution. In the last two hundred years, theories that try to explain the course of history based on a single independent variable, have proliferated. For Marx, that variable is economics, but for Sigmund Freud (1856-1939) it is the sexual impulse and the fear of death. For other authors it is technology or religion[82]. All give convincing arguments and examples, but they cannot be reconciled with the others. The current opinion of sociologists tends to consider that all these historical theories oversimplify the problem. Each one may be true in its own way, in certain circumstances, but history as a whole behaves as a function of many variables.

---

[82] Max Weber (1864-1920), *Die protestantische Ethik und der Geist des Kapitalismus*, 1904-1905.

* * *

Let us now turn to another utopia, enormously influential during the nineteenth and twentieth centuries. This is *scientism*, also called the *doctrine of indefinite progress*, and *scientific utopianism*. We have mentioned *The New Atlantis* by Francis Bacon[83], the first utopia that proposed an ideal society governed by scientists, a view similar to that of Plato's Republic, but adapted to the European Renaissance way of thinking, as it replaces philosophers by their modern equivalents.

From a theoretical point of view, the doctrine of indefinite progress dates back to 18th-century thinkers such as Claude Henri, count of Saint-Simon (1760-1825), who advocated a society governed by technicians, or as Marie Jean Nicolas de Caritat, Marquis de Condorcet (1743-1794), one of the authors of the Encyclopédie and of *Sketch for a historical picture of the progress of the human mind*[84].

The main philosophical support for scientism was given by Isidore Marie Auguste François Xavier Comte (1798-1857), founder of positivism. According to his theories, detailed in his work *The course in positive philosophy*[85], human evolution went through three successive stages: theological, dominated by religion, since the origin of man to the dawn of Greek civilization; metaphysical, dominated by philosophy, up to the sixteenth century; and positive, the era of science, where we are. Comte believes that scientific development will culminate in the total systematization of

---

[83] See note 70 in this chapter.
[84] *Esquisse d'un tableau historique des progrès de l'esprit humain*, published in 1801.
[85] *Cours de Philosophie Positive*, 1830-1842.

knowledge about man and society, the object of the new science of sociology (a name that he coined), which will eventually absorb all the other disciplines of human knowledge.

Biological evolutionism, formulated by Darwin, was misunderstood by the supporters of these theories, who welcomed it as the scientific confirmation of their theories. As a result of this misunderstanding came the doctrine of indefinite progress, popularized in the early twentieth century by Herbert George Wells (1866-1946)[86]. Man, after passing through a primitive stage where he felt overwhelmed by the power of nature, reached a level of evolution where he learns to dominate the environment. Science replaces the obscurities of ancient mythologies in human knowledge. In the future, the increase in knowledge and the mastery of nature will increase, making us happier, freer and more powerful. After many centuries (on the scale of biological evolution, time does not count), we'll become a race of demigods. The supreme enemy, death, will be defeated. Man will have ascended his throne[87].

That science is the panacea that, in the long run, will solve all our problems, was a common belief of many people, or at least of the educated and semi-cult citizens of the West, until well into the twentieth century. Today, however, there is a growing distrust towards science in the abstract, and towards scientists in particular. The suggestion that men of science should govern society, which sixty years ago seemed an inevitable consequence of human evolution, would today be rejected by many. What has happened?

---

[86] *Outline of History*, 1920.
[87] These theories have been recently resurrected by Transhumanism.

The Fifth Level of Evolution

Modern man has discovered that scientific advances are a two-edged weapon. The industrial revolution improved our living conditions, but has also introduced a high degree of pollution, which endangers the survival of life on our planet. Advances in medicine have controlled many terrible plagues of the past, but they have also caused the explosion of population. Chemistry provides us with medicines, plastics and many useful products, but also mustard gas and terrible weapons, plus the plastic pollution of the sea. Atomic fission provides an important source of energy, but makes nuclear bombs possible and introduces the problem of radioactive waste. Biotechnology will let us correct genetic diseases, but also lends itself to human manipulation at a level that belittles the greatest dictator in history. Where is the utopia, our happiness thanks to science?

People of the 21st century understand, hopefully not too late, that science is not the panacea that ensures salvation. Again, we are forced to recognize that material things and knowledge are neither good nor bad in themselves; that everything depends on their use. In short, that man is the sole responsible of the failure of the scientific utopia.

On the other hand, the doctrine of indefinite progress has received a heavy blow as the result of the latest advances in philosophy of history. The big four of the twentieth century in this discipline, Oswald Spengler (1880-1936) in *The Decline of the West*[88], Arnold Joseph Toynbee in *A Study of History*[89], Alfred Louis Kroeber (1876-1960), author of *Configurations of Culture Growth* (1944), and Pitirim A. Sorokin, in his *Social and Cultural Dynamics*[90],

---

[88] *Der Untergang des Abendlandes*, 1923.
[89] See note 67 in this chapter.

coincide in interpreting history, not as a more or less continuous ascending movement, but as a succession of ascents (civilizations) and setbacks (collapses). This does not mean that the course of history is reduced to a superposition of repetitive cycles, always the same, where nothing evolves. Using the simile used by Toynbee, we can compare it rather to a vehicle whose wheels, when turning, make it move forward: *This harmony of two different movements - a major, irreversible one, carried on the wings of a minor one, which repeats itself - is, perhaps, the essence of what we understand by rhythm.*

In other words, there is no indefinite progress. Perhaps in two hundred years, the level of modern science won't be better than it is now, or even it will be worse. Perhaps our civilization will have stagnated or disappeared. But if ever a new civilization emerges from our ashes, it will possibly get further than Western culture, at least in some of its attainments (not necessarily scientific), as our culture surpassed the Greek.

* * *

At the same time as the scientific utopia, another utopia, almost opposite, whose origin dates back from the 18th century, has also gained many followers. Jean Jacques Rousseau (1712-1778), in his works *The social contract*[91] and *Discourse on the origin and basis of inequality among men*[92], formulated the theory that human beings are good and free by nature, but society, opposing their natural state, makes them bad and chains them. The perfect society

---

[90] 1937-1941, in four volumes.
[91] *Du Contrat Social*, 1762.
[92] *Discours sur l'origine et les fondements de l'inégalité parmi les hommes*, 1755.

will be the one that returns to the original situation, goes back to nature and gives us back our freedom and innate goodness. In other words: the wild man was perfect. Thus, Rousseau's utopia looks to the past rather than to the future, as most of the theories we have reviewed so far.

Retrograde utopias (in the etymological sense of the term *retrograde*) have been adopted, since the end of the 20th century, in radical environmental movements, which try to reverse the direction of human evolution and return to an illusory past, which never really existed and has been idealized through a fragmentary knowledge of history. These environmentalists believe that the salvation of humanity depends on forfeiting most of our technological achievements, from which they just perceive their negative effects: damages to the environment, the dangers of nuclear energy... They sing praises to life in communion with nature and to the return to a more primitive way of life.

Retrograde utopias, with their desire to go back to a situation that never existed, are unrealizable. If we follow that path, we'd cause a much worse and certain catastrophe for humanity than those announced by doomsayers. Technological dangers exist, but technology also gives us hope of escaping their dire effects.

* * *

The importance of education as a factor of perfect societies is an obvious consequence of the previous analysis. Both in More's *Utopia*, as in Plato's, as in Marxism, the stability of society is based on the education of its members. Retrograde utopias also consider education very important. Rousseau's educational ideas were described in his book *Emile, or on education*[93], one of the most influential he ever wrote.

Do we have reason to assume that this is true, that education will make us happier, better, more *conformist*? A society made of conformists would be stable and free of internal changes that, by definition, would make it worse. Remember that we are talking about perfect societies.

Apart from the question of whether it would be worth living in a society of conformists, statistics do not confirm the assumption that education is a determining factor in maintaining social stability. *Nor does historical induction authorize the widely accepted opinion that an increase in elementary culture, scientific discoveries, technological inventions or democracy would reduce social antagonisms. From the eighteenth to the twentieth century, schools, scientific discoveries, inventions, population culture and democracy have increased together; in the nineteenth and twentieth centuries they all increased in enormous proportions and, nevertheless, wars, revolutions, conflicts between groups, and crimes show that antagonisms have experienced unprecedented growth... Popular opinion, according to which these factors foster... solidarity or antagonism is not supported by the existing body of evidence*[94].

On the other hand, conformist (therefore stable) societies are quite unattractive. Just consider the two great dystopias of the twentieth century, *Brave new world*[95] and *Nineteen eighty-four*[96]. The

---

[93] *Émile*, 1762.
[94] Pitirim A. Sorokin, *Society, Culture and Personality*, III part, chapter 6.
[95] By Aldous Huxley (1894-1963), 1932.

oppressive feeling that seizes the reader of these two novels is almost unbearable. In both cases, the very rare nonconformists that may arise are excluded from society: in the first, they are banished forever to an island; in the second, more subtle, exclusion is only temporary: the rebels are subjected to brainwashing, with the goal of destroying their spirit and turning them into raw material on which the social planner can act, remodel and educate, until the rebel recovers and is readjusted to society.

The two dystopias are horrible, but they have a power of conviction, a probability, far superior to their opponents, with the possible exception of the Marxist utopia, which was able to drag men and arouse violent feelings, in favor or against. Any society that wishes to perpetuate itself indefinitely at all costs must resort to inhuman and dehumanizing control methods. Any society, made of free, selfish men, inclined to evil, must be unstable by nature, unless its members are forced to adopt a permanently conformist attitude. This is what C.S. Lewis (1898-1963) calls the abolition of man[97].

* * *

In short: the fifth level, the perfect society, although it has not yet come into existence, draws our imagination since millennia. There is no unanimity among various authors about its characteristics, although the different versions of the *utopia* can be grouped in a few families:

- *Rationalists* (we may call them thus) describe a society governed by reason (philosophers, scientists or

---

[96] By George Orwell, pseudonym of Eric Blair (1903-1950), 1949.
[97] *The Abolition of Man*, 1947.

technologists). In this group we can classify Plato's *Republic*, Francis Bacon's *New Atlantis*, and the doctrine of indefinite progress in any of its forms and varieties.

• Communists, who consider private property to be the sole cause of all the evils of today's society, and therefore believe that its abolition would be enough to create a perfect society. Here we can group Thomas More's *Utopia*, *The City of the Sun* by Campanella, as well as Marxism.

• *Retrograde*, who want to go back to an idealized past that only exists in their imagination, such as Rousseau's theories and radical environmental movements.

• Finally, *dystopia*, which does not really try to describe the fifth level, but points to those paths that will take us away from it, emphasizing the negative characteristics of our current society and its possible future extrapolations.

Some of these utopian theories have been put into practice with negative results in every case. A serious sociological analysis, such as that carried out by Sorokin, draws rather pessimistic conclusions. Some of the dystopias, especially *Brave new world* by Huxley, turns out to be much more plausible than any of the utopias, and threatens to become a reality in a future no longer too far away. Education, which all utopian theories make one of the pillars of the perfect society, leaves much to be desired in reality, and does not seem capable of guaranteeing the future stability of the fifth level. If we really consider a desirable objective the perfect society, we must find other solutions and other means to ensure its stability.

9. The Omega point

In August 1913, in a quarry located in the small town of Piltdown, Sussex, England, a young Jesuit priest observes the excavations. Suddenly, he leans down, picks up an object and shows it to his companion. It's a small bone, a tooth: a human canine.

The priest's name is Pierre Teilhard de Chardin. He is thirty-two years old and was ordained two years ago. Born in Sarcenat (France) on May 1, 1881, he is the fourth of the eleven children of a landowner who loves geology, an interest he passed on to his son. At the age of ten, Pierre entered the Jesuit school in Mongré as a boarding pupil, and at eighteen he went to the seminary of Aix-en-Provence. He studied philosophy in the island of Jersey, in the English Channel, specializing in paleontology. At twenty-four he was sent as a teacher of physics and chemistry to the Jesuit College in Cairo, where he remained for three years. Upon his return to Europe, he finished his preparation for the priesthood, received the sacred orders and began his scientific and research activities in the field of his specialty. Finding that canine is his first serious discovery.

His companion is Charles Dawson, a lawyer by profession, an enthusiast of geology and archeology. He serves as secretary of the Archaeological Society of Sussex. In recent months he has become famous in the scientific world of the United Kingdom, in relation to one of the most sensational paleontological findings of recent times: the Piltdown man.

Darwin never said or wrote the famous phrase attributed to him, *man descends from the apes*. In fact, this phrase is false, if it is interpreted, as often happens, as meaning that one of the present species of apes has been the ancestor of man. It can be considered true, however, in the sense that man descends by evolution from other species of living beings, some of which were also ancestors of the anthropoid apes of today. These species, which could be called *intermediates* between man and ape, were therefore called *missing links*. At the time just after the publication of *The Origin of Species*, no trace of them had been discovered.

When Darwin's great work was published in 1859, the first remains of *the Neanderthal man*[98] had been found three years earlier, but nobody knew how to interpret those remains, and the fact that their cranial capacity was on the average greater than ours cast doubt on their possible role as an intermediate between man and ape.

In 1891, the Dutch physician Marie Eugène François Thomas Dubois (1858-1940) discovered on the island of Java fragments of a jaw and a skull, as well as a femur, which seemed to belong to a much more primitive type of man than Neanderthal. The name he assigned, *Pithecanthropus erectus*[99], indicates that his discoverer thought that he had found the famous missing link, although he seems to have recanted later. In fact, to escape the controversy

---

[98] Neanderthal means, in German, *the valley of the Neander River*, near Dusseldorf.
[99] The literal translation of these two words, one Greek, the other Latin, is *ape-man who walks erect*.

caused by his discovery, he refused for years to let anyone examine the bones of *Pithecanthropus*.

The next discovery took place in 1907, in a sand quarry at Mauer, near Heidelberg. It was a very primitive human-looking jaw and provisionally received the scientific name *Homo heidelbergensis*. Piltdown's discovery took place immediately later.

By 1908, a worker at the Piltdown quarry gave Dawson a fragment of an ancient-looking human parietal. Three years later, in the same place, Dawson found a piece of frontal bone. The amateur archaeologist then sought the help of a professional, Arthur Smith Woodward, responsible for Geology at the British Museum. Both spent the summer of 1912 exploring the Piltdown quarry, where they found more remains, including a jaw. The bones of the skull looked modern, but the jaw looked simian. Dawson and Woodward decided that all the pieces belonged to the same skeleton, and in December 1912 they published the discovery, presenting it as a missing link between man and ape. Many British paleontologists accepted this interpretation without question. Perhaps it enhanced their patriotic pride, placing Britain at the level of other countries where similar findings were being made. In fact, Woodward published a book about the Piltdown man with the significant title *The earliest Englishman*, giving the fossil the scientific name *Eoanthropus dawsoni*[100], in honor of his friend.

A few specialists had doubts about the interpretation offered by Dawson and Woodward. In 1913, D. Waterston, a professor at King's College in London, hypothesized that the jaw and the skull

---

[100] From *eos*, dawn. *Eoanthropus* means *the man of dawn*, the man from the beginning of time.

could belong to two different individuals: the first simian, the second human. However, in 1915 Dawson made a new discovery three kilometers from the Piltdown quarry, within the same municipality: two skull fragments and a tooth, the first human-looking, the second simian. The coincidence seemed excessive and almost all researchers accepted the official interpretation. After Dawson's death, which took place in 1916, nothing new was found on Piltdown.

The first suggestion that Piltdown man was a deliberate scientific fraud was proposed by Gerrit Miller Jr. of the Smithsonian Museum in 1930. But it was not until 1953 when Joseph Weiner and Wilfrid Edward Le Gros Clark made a complete study of the remains, using methods such as the analysis of the amount of fluoride and nitrogen in the bones, from which their relative age can be deduced. It was proved without doubt that the set of bone fragments found at Piltdown had been deliberately planted by someone, with the intention of deceiving the scientific community and suggesting that a missing link between man and ape had been found. In fact, the skull bones turned out to belong to modern man, while the jaw had belonged to an orangutan. All the bones had been treated with paints and other procedures, to increase their apparent antiquity. Other fossils, hippo and elephant, which had been found scattered around, came from different countries and from different times. Some of them, apparently, had been stolen from the British Museum.

Since 1953 to the present, the author of the fraud has not been discovered. Dawson is one of the main suspects, of course, but some think that he did not have the necessary preparation to organize such a complex fraud, and he might have been fooled by a professional who wanted to strengthen his theories about the origin of man. Several researchers have proposed different hypotheses, more or less plausible, about the identity of the trickster, who took his secret to the grave. One of these hypotheses blames Pierre Teilhard de Chardin, although others, far more reasonable, point to personalities of the paleontological world better known at that time.

* * *

But let's get back to Teilhard de Chardin, who at the outbreak of World War I volunteered as a stretcher in the medical corps of the French army, although his priesthood allowed him to play the less exposed role of chaplain. His activity, moving wounded soldiers during the terrible battles of that war, won him the Military Medal for acts of courage, as well as the Legion of Honor. After the conflict, Teilhard received a doctorate from the Sorbonne and lectured on Geology at the Catholic Institute in Paris. In 1923 he went to China to participate in a paleontological expedition. Upon his return to France in 1926, he was forbidden teaching for theological reasons, and went back to China, where he remained for almost twenty years. There he participated in the second discovery of his career, much more important than the first, and free of fraudulent connotations: the man from Beijing.

The head of the expedition was the Canadian anthropologist Davidson Black, who in 1927 discovered a human tooth in the cave of Chu-k'u-tien, a town about 50 kilometers southeast of Beijing. From the study of the tooth, Davidson concluded that it was a new species, a possible ancestor of modern man, to which he gave the name *Sinanthropus pekinensis*[101]. Subsequent excavations led to the discovery of skulls, jaws and other bones belonging to more than forty individuals, as well as the remains of fire, indicating that the Peking man knew how to use it. In 1932, Black proposed that the Peking man must be related to *Pithecanthropus erectus*, a theory later confirmed, although *Sinanthropus* turned out to be more recent than *Pithecanthropus*. Both forms are classified today in the species *Homo erectus*.

Teilhard published in the *Revue des Questions Scientifiques* several papers related to his activities in the discovery of *Sinanthropus*. These papers are compiled today in the volume *The appearance of man*[102], which contains his scientific publications related to this topic between 1913 and 1954.

When World War II began, Teilhard was in China, where he had been appointed advisor to the Geological Survey Service, a position he kept until the end of the conflict. The war had dire consequences for human paleontology, as the remains of the Peking man were lost when the ship that transported them to the United States, to put them out of the danger of the Japanese invasion, sank in the Yangtze. Fortunately, subsequent excavations carried out in 1958 in Chu-k'u-tien found new remains.

Upon his return to France in 1946, Teilhard tried to occupy a teaching position in the Collège de France, but gave up in the face of the opposition of his superiors, for the same reasons that he had been forbidden teaching in 1926, about which we'll come back later. In 1951 he moved to New York to work at the *Wenner-Gren Foundation*[103], with which he kept good relations, as this

---

[101] Meaning *the Chinese man from Peking* (Beijing).
[102] *L'apparition de l'homme*, 1956.
[103] Then the *Viking Foundation*.

foundation had sponsored his two paleontological expeditions to South Africa, where the search for human ancestors had led to important findings, parallel to those of the Peking man, about which we'll talk next.

Pierre Teilhard de Chardin died in New York of a heart attack on April 11, 1955.

* * *

Let's go back to the South African discoveries: in 1924, the Australian-South African anthropologist Raymond Arthur Dart discovered in Taung, South Africa, fragments of the skull of an immature individual belonging to an unknown species, a new intermediate link between nonhuman primates and man, whom he baptized with the name *Australopithecus africanus*[104]. This theory generated much controversy, but later excavations confirmed it. Today, of all the ancestors of man, this is the only one that retains its original scientific name, as the others were re-baptized when the Hominid classification was reorganized.

Teilhard de Chardin took part three times in one of the most interesting scientific adventures of the twentieth century: the discoveries of the Piltdown man (which was fraudulent), the Peking man, and the *Australopithecus*. It is curious that he is now less known for his scientific activity than for his philosophical and theological theories. For his part, he always thought that his theories were scientific.

The life and work of Teilhard de Chardin, widely divulged between 1955 and 1973, was so fascinating that it inspired several

---

[104] Meaning *South African Ape*.

fictional characters, based on him. For example, Jean Télémond, in the novel *The shoes of the Fisherman*[105], and Father Merrin, the exorcist[106].

At the beginning of the twentieth century, the theory of evolution had convinced scientists, but was still subject to resistance by religious thought, both Catholic and Protestant, which saw threats in this theory for some basic concepts, such as the monogenetic origin of man and the doctrine of original sin. Shortly before the death of Teilhard de Chardin, in 1950, Pope Pius XII published an encyclical[107] where he addressed evolutionism, declaring it compatible with the doctrine of the Church, provided certain conditions, related to the direct creation by God of the human soul, and with the question of original sin. Later, in the 1990s, Pope John Paul II admitted the scientific foundations of the theory of evolution, maintaining the dogmatic requirements set forth by his predecessor.

Genesis, the first book in the Bible, contains two independent accounts of the origin of man. The first (Gen. 1:26-30), shorter and philosophical, does not give details about the way of creation, nor makes distinctions between man and woman. The second (Gen. 2:4-3:24) is written in mythological style and is the basis for the traditional doctrine of original sin, which can be summarized as follows:

The first human couple was created by God from pre-existing material[108], endowed with *preternatural* properties, such as

---

[105] Morris West, 1963, adapted to the cinema in 1968.
[106] William Peter Blatty, *The exorcist*, 1972, also adapted to the cinema.
[107] *Humani Generis*, meaning *About the human genus*.
[108] Represented in Genesis by *clay* or *the dust from the ground*, with

immortality, and placed in a privileged place, Eden or the Earthly Paradise, where they could live from their work without effort or pain. Unfortunately, when they were subjected to a test, they failed, falling into the temptation of pride (the desire to be like gods).

The failure of the first human couple introduced physical evil (pain and death) and moral evil (sin and bad inclinations) into the world. Since then, all human beings (with two exceptions) have been conceived in original sin, a state of innate rebellion against God that incapacitates us to achieve salvation. To save fallen humanity, the second person of the divine Trinity incarnated in Jesus Christ, who assumed upon himself all the sins of the world and paid for them with his life. From then on, his merits can reach every human being through baptism, which erases original sin and reconciles man with God.

Christ thus appears as Adam's antagonist, for he came to rebuild what Adam had destroyed: *...as by one man sin entered into the world, and death by sin, so death passed onto all men... as by the offense of one, judgment to condemnation came upon all men, even so by the righteousness of one, the free gift unto justification of life came upon all men. For as by the disobedience of one many were made sinners, so by the obedience of one shall many be made righteous.*[109]

Evolutionary theory clashes with the traditional doctrine of original sin in two places. The first is monogenism: in the Christian interpretation of the origin of man, the first human couple must

---

which God formed the first man's body.
[109] Rom. 5:12-21.

have been unique. Only in that way could they represent all mankind, so that their failure would drag us all with their fall. The theory of evolution, in its neo-Darwinian form, prevalent today, argues that natural selection acts on populations rather than individuals, that its action is statistical, so monogenism is not considered plausible.

The second problem refers to the consequences of original sin. For current science, it is evident that pain and death existed on Earth long before the appearance of man, as they are two fundamental properties of life and the very basis of natural selection: the survival of the most apt implies the suffering and statistical elimination of less adapted individuals; death is indispensable in every living system in evolution, because in this way one population leaves room for a slightly different one, made of its descendants. From this perspective, the history of life on Earth would be meaningless in the absence of death and suffering.

A way to escape this contradiction assumed that man, at the time of his appearance, was endowed with privileges over the rest of creation. The first human couple would have been exempt from physical death, pain, fatigue, and so on. These privileges were lost, for themselves and their descendants, as a consequence of the first sin. With this interpretation, the apparition of man would have been a discontinuity in the progress of evolution, while the effect of sin would have been precisely the restoration of the lost continuity. This explanation is possible, but it has against it the principle of parsimony (*Occam's razor*), essential in the scientific method. If the point of discontinuity is eliminated, the history of life is consistent, while, if it is admitted, history loses its global unity.

In traditional Christian cosmology, the first man was adorned with many qualities. He was supposed to be highly intelligent, capable of the immense responsibility he was assigned: he represented the whole future humanity, all his descendants. Modern science imagines the first man with very different qualities: a low brain capacity, perhaps half that of modern man; a rudimentary intelligence, perhaps no language. How can such a being be charged with the responsibility of deciding, not just for himself, but for billions of beings over millions of years?

It should be noted that the origin of the human species from a simian species does not present problems for the Christian interpretation of Genesis. If the reference to clay is taken figuratively to mean *pre-existing matter*, there is no problem in supposing that such matter may have been the body of animals belonging to a pre-human species.

Any theory that wants to be considered in accordance with the Catholic doctrine on original sin must therefore fulfill the following two essential conditions:

a) Creation was originally free of guilt (in a state of grace).

b) As a result of a personal disobedience, creation was stained and lost its original impassivity.

By 1912, Pierre Teilhard de Chardin knew about the theory of evolution through his scientific studies and by reading Henri Bergson[110]. He soon realized the difficulties that this scientific theory posed for traditional theology and tried to find a rational solution that would reconcile his scientific knowledge with his

---

[110] *L'évolution Créatrice*, 1907.

beliefs. By 1920 he had reached conclusions, which he described in two unpublished notes[111], much later published in the collection of articles *Comment je crois* (1969). These notes, sent to the superior of the Jesuits in Rome, were the reason why he was forbidden to teach at the Catholic Institute in Paris.

The solution proposed by Teilhard de Chardin to the problem mentioned above was really controversial, since it did not save the essence of the dogma of original sin and introduced important theological problems. According to that solution, the universe would have been created in a state of initial disintegration, subject from the beginning to a process of evolution that would be based (it couldn't be otherwise) on pain and death. The original sin wouldn't have been, in this interpretation, a personal sin of one or several individuals, but the state of original disaggregation of the world.

This solution was not satisfactory, since it does not meet the two essential conditions mentioned above, which would make it compatible with Catholic doctrine. In Teilhard's interpretation, the universe would have been created from the beginning in a state of guilt, which in no way could be considered a consequence of a personal sin.

Apparently Teilhard never heard about the *Big Bang* theory, first proposed in 1927 by another priest, the Belgian Georges Édouard Lemaître (1894-1966). It is true that this theory was not generally accepted until after Teilhard's death. If he had known it, perhaps he'd have imagined an alternative solution to the problem of

---

[111] *Fall, redemption and geocentrism*, 1920; *Note about some biblical representations of original sin*, 1922.

The Fifth Level of Evolution

original sin: identifying the newly created universe with the biblical Adam, St. Paul's first Adam. At the time of creation, the universe could have been conscious and free. Filled with pride, he wished to be like God, and as a consequence died, started to expand, and his remains began a slow process of evolution that still continues.

With this interpretation, pain and death would be a direct and automatic consequence of original sin for all living beings in the universe, regardless of the planet or galaxy where life has arisen. It is not necessary to assign the first man on Earth a role in the drama, so the problem of monogenism disappears. Also, the first man's rudimentary intelligence doesn't collide with the first Adam's huge responsibility. Evolution thus becomes a second opportunity granted by God to the universe to produce intelligent beings.

This idea, which came to my mind over forty years ago, can perhaps be considered religion-fiction, as a priest once told me. However, it fulfills all the essential conditions of Catholic doctrine, since the universe would have been created in a state of grace and the cause of its loss of innocence and immortality would be a personal sin. Anyway, my position about this is ambiguous: I cannot say that I believe that this really happened, but the fact of having found a coherent and acceptable explanation solves my rational doubts. If there is one, there may be others, and one among them (no matter which) must be the true one.

**

I will summarize Pierre Teilhard de Chardin's cosmological theory, together with his predictions for the future, which are explained in much greater detail in his fundamental work, *The Phenomenon of Man*[112], and in a more condensed way in *Man's Place in Nature*[113]. Both books could not be published during Teilhard's life, as he wasn't authorized by his superiors due to the theological problems mentioned above, although these problems are not evident in these two works. Although Teilhard, following his vow of obedience, accepted the restrictions he had been imposed, he hoped that his work would be published after his death, as was the case.

All his manuscripts were left under the patronage of a general international and a scientific committee, the latter formed by more than thirty members, including names as well known as Henri Breuil, Pierre Grassé, Julian Huxley, Ralph von Koenigswald, Wilfrid Le Gros Clark, Jean Piveteau, George Gaylord Simpson, Arnold J. Toynbee and Miguel Crusafont Pairó. Over seventeen years, these committees published progressively his complete works, until the last collection of his articles, which appeared in 1973.

In Teilhard's vision, consciousness is a basic property of the cosmos, intimately related to complexity. In the beginning of the universe, there was total disintegration and an imperceptible level of consciousness. In a universe made by a soup of elementary particles, atoms or molecules, no conscious activity can be detected. Teilhard, however, asserts that this consciousness exists, even if it falls below our threshold of perception.

---

[112] *Le phénomène humain*, written in 1938, published in 1955.
[113] *Le groupe zoologique humain*, written in 1949, published in 1956.

As soon as the evolution of the universe crossed the threshold of the appearance of life, consciousness became more and more perceptible, in proportion to the complexity of living beings, which grows over time. Although unicellular beings are too simple to be considered conscious in the sense commonly attributed to this term, they are able to respond to certain stimuli with apparently motivated responses, and show inklings of finality. In plants, invertebrates and lower vertebrates, automatic or merely instinctive behavior dominates, but in birds and mammals there are indications of feelings and of conscious motivation. This trend is seen most clearly in primates, especially in anthropoid apes.

With the appearance of man, a threshold is crossed: consciousness clearly dominates instinctive behavior. Before man, the surface of the Earth was covered by a living layer (the *biosphere*[114]); now it becomes covered by a new layer, which Teilhard calls *noosphere*[115], coined by him. He also applies the name *noogenesis* to the progressive movement towards consciousness in the evolution of the universe.

Throughout the evolution of life, we can observe an irresistible tendency towards divergence. Through genetic diversification, the various species of living beings move away from each other, leading to the formation of a family tree where present species are at the ends of branches, much more away from each other than their ancestors who lived, for instance, five hundred million years ago. After man appeared, this process did not stop. In us, evolution invades a new field and becomes cultural evolution, but the same diversification continues to cause branching in the tree, with

---

[114] From the Greek *bios*, life. It means *the sphere of life*.
[115] From the Greek *noos*, intelligence, mind, consciousness.

human cultures and civilizations playing the role of biological species.

However, in recent centuries, man has completely covered the Earth, invaded all continents, even the least fit for life. On the other hand, the apparent reduction of the terrestrial surface, a consequence of the increase in our speed of movement and the improvement of the means of communication, has caused the reversal of branching, which has been replaced by the opposite phenomenon: the progressive convergence of the biology and culture of the dominant species on the Earth[116].

Teilhard extrapolates this convergence towards the future and concludes that, sooner or later, the unification of the human species will be achieved in what he calls *the Omega point*[117]. At that point, all sentient beings in the universe will unite, as components of a single body, in the same way that the cells of the human body come together to form a higher order being. It is not difficult to realize that what Teilhard calls the Omega point is the same with what I have called here *the fifth level*, although when Teilhard wrote biological research was not advanced enough for him to realize that changes of level have occurred several times, apart from the only case known at that time: the union of many cells to form a living organism of a higher level.

In addition to the two books mentioned, where he explains in detail his worldview, from the philosophical and scientific point of view,

---

[116] In current terms, what we call *globalization*.
[117] The letter Omega is the last in the Greek alphabet. With this name, Teilhard means that the Omega point represents the end or culmination of the process of evolution. He has another reason for using that name, but we'll leave that for the last chapter.

the international committee published two others, equally unpublished, of a very different character, as they can be considered mystical literature. They are *Hymn of the universe*[118], which compiles a few short works written over the years, and *The Divine Milieu*[119], a complete work. Apart from *The Phenomenon of Man*, where Teilhard decided to stay on a phenomenological and scientific level, *The Divine Milieu* provides essential data to understand Teilhard's thinking from the point of view of his Christian faith. We miss, perhaps, a third volume that would combine both perspectives in a unique construction.

This absence cannot be palliated, except in part, with the enormous profusion of articles and notes that he published in specialized magazines during his life, or those that remained unpublished, all of which the international committee compiled into eight thematically grouped collections; the two mentioned above (*The appearance of man* and *Comment je crois*), along with six others: *Vision of the past*[120]; *The future of man*[121]; *Human energy*[122]; *Activation of energy*[123]; *Science and Christianity*[124]; and *Toward the future*[125].

---

[118] *Hymne de l'univers*, published in 1961.
[119] *Le milieu divin*, written in 1927, published in 1957.
[120] *La vision du passé*, 1957.
[121] *L'avenir de l'homme*, 1959.
[122] *L'énergie humaine*, 1962.
[123] *L'activation de l'énergie*, 1963.
[124] *Science et Christ*, 1965.
[125] *Les directions de l'avenir*, 1973.

Manuel Alfonseca

10. Internet as a nervous system

Living beings at levels higher than the first must set up some procedure, so that the lower order units living inside them can communicate with each other. This seems obvious, but let's check:

Inside a prokaryotic cell (a second level organism) numerous nucleic acids coexist[126]: DNA, messenger RNA, transfer RNA... One or several plasmids[127] can also be there, living in symbiosis. All these molecules coordinate with each other through a complex system of enzymes and other substances, which manage gene activation and the triggering of protein synthesis. To transmit internal communications, the cell uses chemical diffusion in the aqueous medium of its protoplasm. Keep in mind that the diameter of a prokaryotic cell does not usually exceed a thousandth of a millimeter (one micrometer). The transmitted molecules are usually measured in nanometers (millionths of a millimeter), so the difference in size between the transmitted objects and the distance to travel is usually not very large.

A eukaryotic cell (a third level organism) contains several prokaryotic cells: mitochondria and possibly chloroplasts. During reproduction, carried out by mitosis[128], all these organisms are coordinated, so that at the end of the process the two daughter cells have a complete endowment of cellular organelles. Although

---

[126] See chapter 3.
[127] See chapter 2.
[128] See chapter 4.

eukaryotic cells tend to be larger than prokaryotes, they are still not visible to the naked eye, and resort to chemical diffusion to transmit information, although this may be aided, during reproduction, by some form of mechanical transmission through the achromatic spindle, a structure that appears only at that time, and is an essential part of the mitosis control system.

Multicellular plants, organisms of the fourth level, frequently need to transmit information, both inside and outside themselves. Those that live on solid ground can emit visual information (colors) or use chemical diffusion through the air, to communicate with other beings of the fourth level. Many plants make use of insects to pollinate, and give off aromas to attract them. The message conveyed is simple: *I am here*. Sometimes, the process is more sophisticated: the orchid *Ophrys sphegodes* releases a chemical that resembles the smell of a pheromone from the female bee *Andrena nigroaenea*, which attracts males of that bee. However, once pollinated, the flower changes odor and gives off a different substance, *farnesyl hexanoate*, which repels males. It is curious that the females of that bee, once fertilized, also release the same substance to get rid of the males.

There are many other examples. When the leaf of a tobacco plant is eaten by a caterpillar, it emits chemicals that attract predators of the caterpillar. Many trees give off various volatile organic compounds, which sometimes report some cause of stress. Chemical diffusion in aqueous medium is also used to transmit information inside the plant between organs or cells far apart from each other, through the system of vessels distributing the sap, by means of physical procedures such as capillarity or osmosis. Sometimes more complex procedures are used, such as the one regulating the coordinated opening and closing of pores in the leaves, which has been compared with distributed computing.

Animals also use chemical diffusion. The endocrine system consists of glands that secrete various substances, transmitted through the circulatory system, which carry chemical signals to distant organs. But the way of life of an animal is usually much more active than that of a plant, so the slowness of chemical diffusion is inappropriate when the distance the signals must travel is large and the response time should be small. Therefore, almost all types of animal organization (except sponges) have some form of nervous system, more or less sophisticated, using electrical signals, which are transmitted much faster than chemical signals.

In the simplest types of organization, the nervous system is decentralized, but in animals of some complexity there is a control center (the brain), located in a special part of the body (the head), whose size is correlated with the complexity of the animal's behavior. There are studies on this complex correlation between the intelligence of living beings and the size of their brain. At first glance, what matters, rather than the size of the brain, is the ratio of its mass to the mass of the body. It is logical: a large animal needs a large brain just to control its own body. Table 10.1 shows the value of that ratio for several species of living beings[129].

---

[129] Data taken from the book *The dragons of Eden*, by Carl Sagan, 1977.

Species	Brain mass (g)	Body mass (kg)	% brain mass / body mass
Raven	10	0.3	3.33
Homo sapiens	1400	60	2.33
Pigmy shrew	0.1	0.005	2.00
Golden carp	0.35	0.02	1.75
Homo erectus	1000	60	1.67
Australopithecus	500	35	1.43
Homo habilis	800	60	1.33
Dolphin	1600	150	1.07
Hummingbird	0.1	0.01	1.00
Baboon	180	21	0.86
Rat	2.5	0.3	0.83
Chimpanzee	350	60	0.58
Coelurosaurus	150	30	0.50
Gorilla	600	300	0.20
Lion	250	220	0.11
Elephant	4500	8000	0.056
Ostrich	50	110	0.045
Cachalot	9000	50,000	0.018
Eel	0.5	4	0.0125
Alligator	15	210	0.0071
Tyrannosaurus	200	10,000	0.0020
Diplodocus	60	20,000	0.0003

Table 10.1. Ratio from brain mass to body mass

Looking at the table, we can see a few anomalies: at the top, with the maximum percentage of brain mass with respect to body mass, is the raven. In fact this is one of the most intelligent birds, but it does not seem reasonable to consider it superior in intelligence, not just to man, but even to the chimpanzee or the gorilla, which appear further down. Similarly, the pygmy shrew and the golden carp occupy the third and fourth places, immediately after modern man, but before all other primates. On the other hand, large dinosaurs have a reputation for having little intelligence, but are we willing to accept that it was significantly lower to that of an eel? Finally, *Coelurosauria* are traditionally considered the most intelligent of dinosaurs; but higher than the gorilla and the elephant?

It can be seen that the intelligence of large animals is undervalued in Table 10.1, while that of smaller ones is exaggerated. It is obvious that the percentage of brain mass with respect to the body mass is not a good measure. There are reasons: apart from the processes usually classified as intelligent, the brain devotes much of its activity to control the functioning of the body, where surface supervision plays a fundamental role, as the surface is the interface between a living being and its surroundings, and contains the sensory organs, which let it interact with its outside.

As the size of the body grows, volume increases much faster than surface (volume increases as the cube of the dimensions; surface grows as the square). For instance, an animal whose dimensions are double, compared to another, has a volume eight times larger, but an area just four times larger. As the brain grows in volume as a function of the cube, larger animals must devote a smaller percentage of their brain to the control of their body surface, compared to smaller animals. Therefore, with the same percentage of brain mass compared to body mass, a larger animal will release part of its brain for intelligent activities, while a smaller one must use almost all of it to control its body.

To compensate for this, another ratio can be used: the logarithm[130] of brain mass to the logarithm of body mass, which has the desired effect. Compared to the simple percentage of brain mass, it assigns a greater value to larger animals and a smaller value to smaller ones. Table 10.2 presents the results, ordered again from highest to lowest. Perhaps in this case the tiniest animals are unfairly treated. The rat has dropped too many places; hummingbirds and pygmy shrews, the smallest of all, have gone down to the last places; but the order of other species seems much more reasonable than in table 10.1. Other measures have been proposed to relate the brain size of living beings with their intelligence, but I don't think it's necessary to detail them further.

---

[130] The logarithm of a number is the exponent we must apply to another number (usually 10) to obtain the first number. For instance, the logarithm of 10 is 1, as $10^1=10$; the logarithm of 100 is 2, for $10^2=100$; the logarithm of 1000 is 3, as $10^3=1000$; etcetera. Notice that logarithms grow much more slowly than the corresponding numbers.

Species	log brain mass (g) / log body mass (g)
Homo sapiens	0.66
Homo erectus	0.63
Dolphin	0.62
Homo habilis	0.61
Australopithecus	0.59
Chimpanzee	0.53
Elephant	0.53
Baboon	0.52
Gorilla	0.51
Cachalot	0.51
Coelurosaurus	0.49
Lion	0.45
Raven	0.40
Ostrich	0.34
Tyrannosaurus	0.33
Diplodocus	0.24
Alligator	0.22
Rat	0.16
Eel	-0.08
Golden carp	-0.35
Hummingbird	-1.00
Pigmy shrew	-1.43

Table 10.2. Logarithmic mass ratio

If someone thinks that, according to table 10.2, even if we occupy the first place, we are not much smarter than the dolphin or the chimpanzee, they are wrong. Like every measurement, also this has flaws (one has just been pointed out). Note that the computation depends only on brain mass. Its distribution is not taken into account. The human brain is not a sphere, it is extraordinarily convoluted, full of cracks, more than the brain of any other animal species, which greatly increase its surface. There are properties of the brain that depend on the surface rather than the volume. None of this has been taken into account in table 10.2. But the fact that, despite all the simplifications, modern man occupies the first place, is quite suggestive.

* * *

Another interesting measurement, related to the previous one, is the amount of information each body contains or can handle, for every living species. In beings belonging to the first three levels of life (viruses, bacteria, algae, protozoa...) genetic information is stored in nucleic acids. Animals, equipped with a nervous system, also have the ability to store information in the nerve cells of their brain, which provides them with a memory and a calculation capacity independent of genetic information. Finally, modern man, the only species capable of evolving at the cultural level, has new procedures for storing information outside his body, such as books and computers.

Table 10.3, whose data have been obtained partly from Carl Sagan's book and partly calculated by me, shows an approximate measure of the average information that each group of living beings can use. Note that, since reptiles appeared, the brain's storage capacity exceeds the genetic information, which for mammals is negligible. In man, on the other hand, extracorporeal cultural information dominates.

Living being	Genetic information	Brain information	Cultural information
Virus	10 a 50 kbit		
Bacteria	1 a 10 Mbit		
Unicellular eukaryote	25 Mbit		
Nematodes	200 Mbit		
Arabidopsis (a plant)	250 Mbit		
Insects	360 Mbit		
Poplar	960 Mbit		
Amphibians	2 Gbit	10 kbit	
Reptiles	3 Gbit	10 Gbit	
Mammals	5 Gbit	200 Gbit	
Man	6 Gbit	10 Tbit	10000 Tbit

Table 10.3. Amount of information in living beings
1 kbit(kilobit)=1000 bit; 1 Mbit(Megabit)=1000 kbit;
1 Gbit(Gigabit)=1000 Mbit; 1 Tbit(Terabit)=1000 Gbit.

The calculation of genetic information is difficult. On the one hand, as we'll see in Chapter 12, we still cannot distinguish the useful information in genomes, and the unusable part of DNA (garbage DNA, as experts call it). If the second is large, the information contained in our chromosomes would be smaller. On the other hand, it seems that a single gene can encode several proteins by using various mechanisms, such as shuffling parts of the genes, or applying chemical corrections after protein synthesis, which would force us to increase our calculations about the amount of information. The figures in table 10.3 should be considered as merely indicative.

Figure 10.1 presents, in graphical form, the evolution of the amount of information handled by various types of living beings over time. The horizontal axis is linear and presents the course of time, in billions of years, from the origin of life (which took place approximately four billion years ago), until the moment of the appearance of the species or group of species considered. The vertical axis is logarithmic (grows exponentially) and shows the number of bits available to the most advanced species in that time (the species with more information). The dark curve represents the sum of genetic, neural and cultural information. The light grey curve corresponds to genetic information only. Cultural information is null for all species, except man. At first, all the information is genetic, so at the beginning the dark and light curves coincide. Later, after animals appeared, about 600 million years ago (3400 million years after the onset of life) genetic information coexists with information stored in the nervous system. Finally, with man, cultural information appears, which quickly surpasses the other two: in fact, all the information we have in the form of books and other external storage media is far superior to what fits in a human brain.

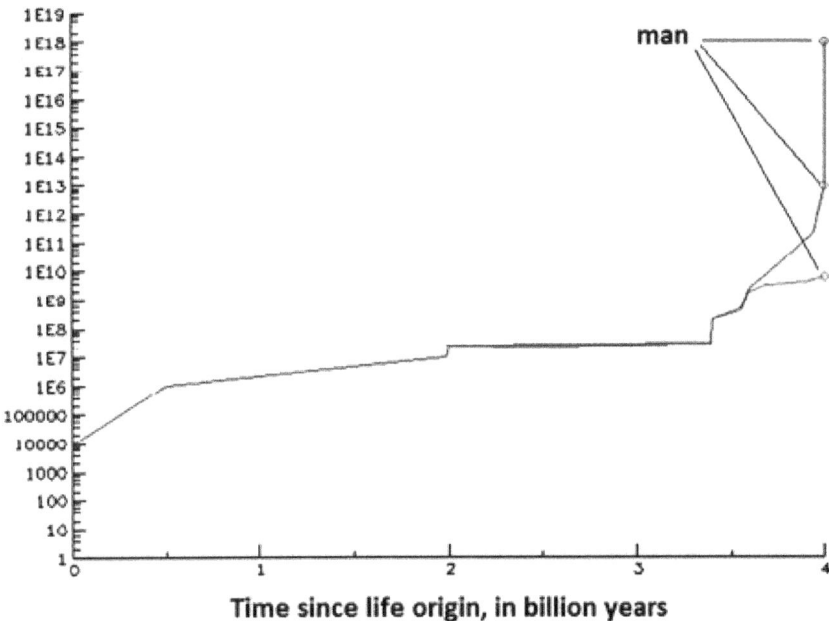

Figure 10.1. Information managed by living being since the origin of life.

It is evident in the previous curve that, throughout the history of life, the amount of information available to the most advanced species of living beings has steadily increased. We saw in Chapter 6 that modern biologists often claim that evolution shows no trends, that there are no species superior to others, that all living things are equivalent. Figure 10.1 provides a new argument to refute it. Quantitative data show, not only that there is a trend in the evolutionary process (towards species capable of processing increasing amounts of information), but also that the human species occupies a privileged place: at this moment, it is the peak of evolution.

Note that at no time have I argued that there can be no setbacks (as may happen to a parasitic species) or that evolution has taken place following a pre-established plan, without any intervention of chance. Remember that the above figures represent average or maximum figures. They are, therefore, statistical data, reducing to a single point collectivities in which there is some variability.

The theory that states that evolution is directed in its smallest details is called *orthogenesis*[131] and was proposed in the 19th century, shortly after Darwin published his theories, by the Swiss biologist Karl Wilhelm von Nägeli (1817-1891), the same who rejected Mendel's discovery. He based his theory on the fossil record, which then seemed to show what looked like continuous series of successive species, as in the horse and the elephant. Later research found that the action of evolution, rather than a bundle of straight lines, resembles a hugely branched tree. Those series turned out to be an artifact introduced by the enormous shortage of fossil remains that had been found by that time.

Unfortunately, Teilhard de Chardin chose the name *orthogenesis* to mean evolution towards always more consciousness, which corresponds to the curve of Figure 10.1 and whose existence can be verified. But the use of a discredited term acted against his theory. It is true that he says he uses it *in the purely etymological and general sense of the term*[132], different from the use made by von Nägeli, but the coincidence must cause confusion.

* * *

---

[131] From the Greek *orthos*, straight, *genea*, origin; orthogenesis thus means *origin in a straight line*.
[132] *Man's Place in Nature*, IV part, *Formation of the noosphere*.

The Fifth Level of Evolution

The incipient societies of the fifth level, polyp colonies, anthills, bee-hives, wasp nests and termite mounds, have not developed anything equivalent to the nervous system of animals. Given the state of standstill evolution in which they seem to have been for millions of years, it doesn't seem probable that they will develop it in the future. To establish communication between members, they resort to chemical means. We have mentioned that sexually active individuals can secrete hormones that spread through the colony and inhibit the appearance of others of the same caste[133]. Ants that find a food source leave a chemical trail when they return to the anthill, which tells other members of the same anthill the direction they must follow to find it. Collecting bees, on the other hand, resort to a very sophisticated dance[134], where the speed and angle of their movements symbolically represent the direction to follow and the distance to travel to find nectar. The dance of the bees is amazing, but requires physical contact between the members of the hive (it takes place in the dark), so its range is smaller than the chemical message of the ants. In both cases, a lot of time is lost between the discovery to be communicated and the reception of the news by other members of the society.

Let's finally tackle human society, another emerging organism of the fifth level. Do we have in this case something equivalent to a nervous system? We do have it, we have just built it, in the last few years.

---

[133] See chapter 7.
[134] Karl von Frisch, *The dancing bees*. Von Frisch (1886-1982) was awarded in 1973 the Nobel Prize in Physiology and Medicine for his studies on the behavior of bees.

For most of history, human beings could only communicate at close range and by primitive means: screams, whistles, drum sounds, fires, torches, smoke signals, flags... In 1588 the arrival of the Spanish Navy was communicated in England by means of bonfires. For long-distance messages, a running man (as in the battle of Marathon), horses or carrier pigeons were used. The messages could be oral, written (since the invention of writing, about 5000 years ago) or coded in some other way.

Sophisticated communication systems were invented a little over two centuries ago. The first message through visual telegraphy, with the system invented by Claude Chappe (1763-1805), was sent on August 15, 1794 and informed the French revolutionary government of the conquest of Le Quesnoy. But the visual telegraph presented important problems: it only worked during the day and in good atmospheric conditions. In just half a century, it was supplanted by the electric telegraph, in whose development participated many inventors: Gauss, Weber, Henry, Cooke, Wheatstone, Morse... Table 10.4 presents a few dates and shows that in just three decades, a truly global rapid communications system was built for the first time in history.

1838	Samuel Morse invents his famous code
24/5/1844	First message sent by Morse in the Washington-Baltimore line[135]
1846	Cooke y Wheatstone create the Electric Telegraph Company
1851	First submarine cable across the English Channel
1862	240,000 km telegraph network in the world
1866	First transatlantic cable
1872	First message Australia-London

Table 10.4. Ephemeris of the electro-dynamic telegraph

The communications revolution then accelerated. First came the telephone (Innocenzo Manzetti, 1850; Antonio Meucci, 1871; Alexander Graham Bell, 1876; Elisha Gray, 1876); then the radio, initially called wireless telegraphy (Heinrich Hertz, 1885; Guglielmo Marconi, 1895; Karl Ferdinand Braun, 1899; Reginald Aubrey Fessenden, 1906; Edwin Howard Armstrong, 1912-33); then television (Vladimir Kosma Zworykin, 1923-28; John Logie Baird, 1925-28); then communications satellites (the first, Echo I, launched in 1960); then personal mobile phones (since 1973) and smartphones (in the 2000s).

At the same time this was happening, the ease with which human beings could move from one point to another on the Earth was also growing. Traditionally, since remote antiquity, travel had been carried out on land by means of animal traction (the fastest was the horse), and in the water by sailboats (pushed by the wind) or by

---

[135] The Biblical quote *What hath God wrought!* (Núm.23:23, KJV) was transmitted.

human traction (rows and oars). Since the 19th century, with the industrial revolution, transport of human beings experimented a revolution. First came the railroad and the steamboat, then, in the twentieth century, air transport. But although we can transfer messages and transport people, this situation cannot be compared to a nervous system: it rather resembles a circulatory system.

Both factors, increasing the ability to move and the ease of communications, have enlarged the reach of every human being (the maximum distance in which one can influence others) until it is now equal to the entire Earth. Teilhard de Chardin thinks that this causes a pressure that tends to unite us, which makes human society converge towards the *Omega point*[136].

On the other hand, the worldwide telegraphic network and its successors cannot be considered equivalent to the nervous system of a living being of the fourth level; at most, it is similar to a bundle of nerves. There was no processing capacity, no memory, apart from the brains of the human cells making society, who just use the networks to communicate. However, in the mid-twentieth century *electronic computers* were invented, and since the seventies they started to be connected with each other through the telephone network, which in addition to voice can also transmit encrypted data.

Do not think that I am comparing a computer with our brain. *Artificial intelligence* comparable to human, which has been much talked since John McCarthy invented the term in 1956, does not exist. It has been said ironically that *artificial intelligence is what*

---

[136] *The phenomenon of man*, book four, chapter I.1.A, *Forced coalescence*.

The Fifth Level of Evolution

we still don't know how to do with a computer. As soon as we manage to solve some of the classic problems classified as intelligent, such as playing chess, they no longer seem intelligent. In 1956 it was predicted that in ten years there would be programs capable of winning against the world chess champion. The prediction was fulfilled thirty years late, in the nineties, and the program that did it (*Deep Blue*, by IBM) owed its ability, rather than to the intelligence of its predictions (that's what the best human players do), to the fact that computers were already so powerful and fast, that they could analyze lots of moves and choose the best one, with a not too complex algorithm.

But I'm not trying to compare electronic brains (as they were called at first) with our brains. We are talking about the creation of an incipient nervous system for an incipient being of the fifth level. We don't need to compare it with the most advanced brain of the fourth level, but with the simplest ones. If we take as reference the nervous system of an annelid, even of an arthropod, our computers don't get so badly. In fact, it would suffice to compare a computer with one of the ganglia of these animals, whose nervous system is decentralized. The parallel is greater than it may seem at first glance: since a few decades, computers can connect with each other and form similar networks, although more complex than those of invertebrates.

At the end of the sixties, telephone lines were used to connect remote terminal stations without computing capacity to the computers of that time, which were much larger and far less powerful than current ones. Each computer was practically isolated, with no direct communication with other computers.

During the mid-1970s, the first protocols making possible communication between computers were invented. Little by little, the first networks appeared: Arpanet (of the United States Department of Defense), Vnet (of the IBM company), Bitnet (the North American university network; the European Union first formed its own network, EARN, and finally connected it to Bitnet), and the Internet (initially created by private companies). All these networks joined one another through gateways, until during the nineties they became integrated into a single network, which retained the name of one of them: Internet.

The next breakthrough took place in 1990, when Tim Berners-Lee and Robert Cailliau, who worked at the European Council for Nuclear Research (CERN), developed a new protocol (*http*) that facilitates communication between computers in Internet. Three years later, CERN relinquished the http protocol to the public domain, so that everyone could use it without paying rights. The response was immediate: that same year, Marc Andreessen and other researchers created the first practical Internet browser, *Mosaic*. A year later, Andreessen and Jim Clark improved the browser and founded a company with the same name as the new browser: *Netscape*. In 1996, Microsoft launched its own browser: *Explorer*. With the help of these tools, and essentially relying on private initiative, in a very short time the *World Wide Web* was formed, which lets anyone to access easily, often free of charge, documentation, information and files distributed through millions of computers located anywhere in the world.

In a story published in 1963, titled *Dial F for Frankenstein*, Arthur C. Clarke, a British science fiction engineer and writer, predicted that when all the computers on the Earth were connected via satellite, they would take control of the planet, snatching it from the human species. Fortunately, this has not happened. Clarke was more successful in another prediction[137], global communication through geostationary satellites, which he foresaw fifteen years before it was actually implemented.

It is curious that Clarke published, in the early sixties, a serious prediction of all the scientific advances that, in his opinion, were to take place, every decade, from 1970 to the year 2100. Of all his many predictions for the six decades now in the past, Clarke just guessed two: the landing on the Moon (which was very predictable, given the American space plans) and the mobile phone, which he called *individual radio*. As an expert radar engineer, Clarke's two great predictions belong to his own field.

With the global web, a true nervous system common to all mankind begins to appear. The information contained in the nodes of the network is surprisingly large. For the first time, the set of cables joining us has a memory. It also has a large computing capacity, although it is distributed: there is nothing so far that corresponds to the brain of a vertebrate. The being we are building has no head.

Like any *living* system, our incipient nervous system is subject to diseases and parasites. There is a harmful spread of viruses, worms, Trojans and other harmful species through the World Wide Web. There are millions of attempts a year, including those carried out by specific individuals or groups (*hackers*, *crackers* and other varieties), against the security of computers or to collapse the use of certain services. These activities, carried out with criminal

---

[137] *Extra-terrestrial relays*, published in *Wireless World*, 1945.

intentions or simply with the wish to annoy others, can be compared with the behavior of cancer cells in living beings of the fourth level.

Naturally, it is not enough that the global network contains a great deal of information: it must be possible to use it when convenient. Therefore, the next step after browsers were search engines. First *Wandex*, then *Yahoo*, *Altavista*, *Google*, have become essential tools, which many of us use every day.

One might think that a nervous system formed by innumerable computer centers connected to each other through telephone lines would be unmanageable. The reality is quite different. The connectivity of the global network is very complex, there are nodes with many connections, and connections between very distant nodes. This results in what has been called the *small world effect*.

Suppose we want to express graphically the relationship between two people. If each person is represented by a circle, and the relationship between them by a line joining the corresponding circles, we get a *graph*, a tool widely used in mathematics and technology. Since there are more than seven billion people in the world, the graph will contain about seven billion circles and a much larger number of lines.

This graph has a peculiar structure: most lines join circles close to each other, since most of our acquaintances do not live very far from us, but from time to time there are lines joining circles far apart, because many people know someone located on the other side of the world. In the sixties, the American sociologist Stanley Milgram discovered that, in graphs with this structure, the maximum number of degrees of separation between any pair of nodes is about six. That's why he called them *small world graphs*, as it's easy that any two people who meet casually will discover that they have some common acquaintance (or that both know someone who knows the same person). Lots of systems representable with graphs exhibit similar properties.

The World Wide Web can also be represented by a graph. Computers will be circles. We'll draw a line between two circles when there is a telephone link connecting them. This graph also has the *small world* properties, which means that any two computers can connect to each other by short paths. This makes the network very flexible and facilitates access to information, whatever country in the world you are in.

* * *

Assume that the previous analysis is correct. The incipient being of the fifth level that we are building on Earth would have a body made by human beings, and a primitive nervous system, consisting of several billion computers, connected to each other by the World Wide Web. *But what kind of fifth level being are we talking about? What unnatural mix is this: living beings on the one side (us), and machines on the other? Where are we going?*

True, at first glance, what we are describing recalls the famous cyborg[138], an hypothetical being, typical in science fiction, the result of the implantation in man of automatic electronic devices that control its biological functions, as well as artificial senses that facilitate life in environments far removed from ours: for instance, in the sidereal spaces.

---

[138] A contraction of *cybernetic organism*.

Yes, but this is quite different. Despite being a man-machine symbiosis, a cyborg is still a being in the fourth level. Here we are talking about a higher order being, where each human individual is equivalent to a cell, while our electronic computers build an incipient nervous system, external to each of us, but internal for the incipient being of the fifth level. A being baptized by Fernando Sáez Vacas with the name *Homo noosferensis*[139], clearly influenced by Teilhard de Chardin.

Note that, in this context, the question of whether artificial intelligence is possible (in the style of classic science fiction robots) ceases to matter. The development of a nervous system for the being of the fifth level is much more important. Remember that what we have built so far is at the level of an annelid, with a decentralized nervous system, a headless body. There is a long way to go.

Anyway, biology has not yet said its last word. The twentieth century was not just the era of information and communication technologies. In its last decades, it also led to the emergence of a new form of engineering: *biotechnology*. In the future, a joint evolution of our body and our instruments could take place. In the next two chapters we'll try to give an idea of the current situation of biotechnology.

---

[139] *Más allá de Internet: la red universal digital*, Centro de Estudios Ramón Areces, 2004.

11. Must we renounce reproduction?

As we saw in previous chapters, the transition from one level to another has taken place at least three times in the history of life:

- From nucleic acids (first level) to the prokaryotic cell (second level).
- From prokaryotic cells (second level) to the eukaryotic cell (third level).
- From eukaryotic cells (third level) to multicellular beings (fourth level).

To this we must add that the transition from the second to the third level probably happened twice independently:

1. A prokaryotic cell learned to live inside another prokaryote, transforming itself into mitochondria.

2. A prokaryotic cell learned to live within one of the cells in the previous group, transforming itself into chloroplasts.

All eukaryotic cells have mitochondria, while not all have chloroplasts, so it seems likely that the transition occurred in two phases, as indicated.

Similarly, the transition from the third to the fourth level must have taken place several times independently. There are some groups whose multicellular organisms have no differentiated tissues, and therefore can be considered incipient beings of the fourth level: *Acrasiomycota* (molds), *Rhodophyta* (red algae), *Chrysophyta* (golden algae), *Phaeophyta* (brown algae), *Labyrinthulida*, *Oomycota* and *Xanthophyta*. These groups sometimes include unicellular and multicellular forms, even both in the same species.

The three great traditional kingdoms, fungi, plants and animals, may also have reached multi-cellularity by several independent paths. Among fungi, for example, *basidiomycetes* are all multicellular, while *ascomycetes* (a group that includes yeasts) have multicellular and unicellular forms. The same goes for plants, which could have reached the fourth level by up to five different paths, and animals, which would have reached multi-cellularity in three different ways: sponges; *placozoa* (or *mesozoa*); and metazoans proper (all the other types of organization).

Finally, we have a few examples of incipient transitions to the fifth level, which have also occurred independently:

- At least twice for *Coelenterata* (*Radiata*): coral polyps and Siphonophorae.
- Four times for insects: termites, bees, wasps and ants.

Level transitions turn out to be a relatively frequent phenomenon in the history of life, as they seem to have happened independently more than twenty times. It is true that this phenomenon is less frequent than the appearance of a new species, as it is estimated that throughout the history of the Earth about one hundred million different species have appeared. This figure includes both present (known or unknown) and extinct species.

We have a pretty clear idea of how the appearance of new species takes place. According to the Neo-Darwinist synthetic theory, one species can branch into two species, its descendants, by a process

The Fifth Level of Evolution

of divergent evolution. Sometimes a population is divided into two isolated groups as a result of a new geographical barrier (a mountain range, a sea invasion, etc.) that prevents genetic exchange between the individuals of both populations. The two groups are then subjected to slightly different environmental conditions, and the action of natural selection will diversify their genetic makeup. If the separation is in effect for many generations, different species and even different genus can be formed. This is what happened with the famous finches of the Galapagos Islands that gave Darwin the idea of the origin of species. The birds in each island were isolated from their kin in other islands of the archipelago and gave rise to different species, adapted to live in slightly different conditions.

Things are not so simple when we must explain the process of uniting several lower level beings to form a single higher level being. So far, there are no theories explaining what happens in these cases, acceptable for all specialists. The main problem can be stated as follows:

Natural selection favors the statistical survival of those individuals best adapted to the environment, for in the long run these will leave a greater number of descendants. Therefore, different individuals of a species not only compete with those of other species that want to occupy the same ecological niche, but also with those of the same species. Consequently, natural selection must favor selfishness, which should become the fundamental law for living beings. In fact, this happens very frequently.

It is true that, in some living species, generally those in which consciousness reaches its maximum development, altruistic behaviors also appear. It has been pointed out that, when a leopard threatens a troop of baboons, some individuals, generally chosen among the strongest, face the predator and give the rest of the troop time to escape, thus risking their lives to save their companions. The explanation of this type of behavior was a problem for evolutionary biologists of the mid-twentieth century, who developed complicated theories, based on the principle of the *selfish gene*[140]. Thus, for instance, it was asserted that an individual would sacrifice itself to ensure the survival of two of its children (each of whom shares 50% of its genes) or four grandchildren (each of whom has 25% genes in common), or a different number, depending on the degree of kinship: siblings, cousins and other relations.

Things get more complicated when we try to explain a change in level, like those we have been describing. In these cases, several individuals of the lower level, who are going to form a single higher level individual, must completely renounce selfishness and permanently adopt a totally altruistic behavior, for the higher order individual could not survive if each of its "cells" is seeking its own and exclusive benefit.

Some researchers[141] have concluded that a change in level (or, as they call it, a *metasystem transition*) may not be viable unless all individuals who come together to form a new higher-level organism give up their ability to reproduce, except just one, or very

---

[140] See note 54.
[141] See Francis Heylighen & Donald T. Campbell, *Selection of organization at the social level: obstacles and facilitators of metasystem transitions*, in *World futures: the Journal of General Evolution*, vol. 45, p. 181-212, 1995.

few. Only in this way natural selection would cease to act, since individuals who do not reproduce don't compete, and altruistic behavior could develop.

In fact, it's not difficult to find examples associated with level transitions where this condition was actually met. Let's look at a few:

- When several first level nucleic acids joined together to constitute a second level being (a prokaryotic cell), all the different forms of RNA renounced the ability to reproduce on their own, which only chromosomal DNA conserves.

- When several prokaryotic cells joined together to give rise to an eukaryotic cell, ribosomes, mitochondria and chloroplasts gave up their independent reproduction, leaving it under the control of the host cell, whose genetic material is contained in the nucleus.

- When unicellular beings evolved to become multicellular, a few, the most primitive (such as certain algae), kept the reproductive capacity of every component, so that a piece torn from one of these individuals, including a single cell, can reproduce until a complete individual is regenerated. Some of this can be observed in more evolved animals, such as echinoderms: if a piece of arm is torn from a starfish, the mutilated individual can regenerate a new complete arm, but the torn piece is also able to regenerate the remainder of the star, so in the end we have two different individuals where at the beginning there was only one. Here we have a traumatic way of reproducing. This is similar to the ability of some plants to proliferate through cuttings.

In almost all animals and many plants, not all parts of the body are able to reproduce. By specializing, most cells give up long-term reproduction. A few (adult stem cells) retain it during the whole life of a multicellular individual, but they are not capable, on their own, to generate another individual. Just a few, so-called *germ cells* or *gametes*, maintain indefinitely their ability to reproduce and generate new individuals, because they have specialized in this vital activity.

- In coral polyps and *Siphonophorae*, some of the individuals joined in a colony can reproduce, while others, specialized in other types of activities (such as digestion, or the production of stinging substances to defend the colony), have renounced reproduction.

- In an insect society (an anthill, a termite mound, a bee-hive or a wasp nest) specialization has reached the utmost. Most individuals who are part of those incipient fifth-level beings have renounced reproduction, except the males and the queen, who dedicates its whole life to laying eggs and sometimes cannot even feed itself. The rest of the members of the colony (workers, soldiers, etc.) are neutral beings, usually unable to reproduce.

***

Let us now look at human society, which we asserted in Chapter 7 is on its way to the fifth level. It is evident that human relations show all kinds of examples of selfishness and altruism. Moreover, in most human acts, it is almost impossible to distinguish between both. Any one of us can verify this, if we examine our deep motivations without trying to deceive ourselves. In one of my novels[142] I explained it in more detail. In chapter 18 one of the characters, Marcius Luculus, says to Flavius Aeolius, the protagonist:

Man is a strange mixture of selfishness and altruism. Whenever we do something good, if you look carefully inside yourself you'll find good and despicable reasons at the same time. It's impossible to separate them. But don't let this depress you. You shouldn't look just at the latter, forgetting the first. That would be the biggest mistake you could make. Accept yourself as you are, learn to live with yourself.

In other words: two trends are inextricably mixed in our behavior. The first is selfishness, which Catholic theology considers the effect of original sin, tries to keep us in the fourth level, to prevent us from reaching the fifth. Natural selection has a tendency to favor this type of motivation, since the members of the human species have not renounced reproduction. The second, altruistic tendencies, which open the way towards the fifth level, must go against natural selection and the basic trend of evolution, both biological and cultural.

We are therefore faced with a dilemma. Must we give up reproduction, so as to reach the fifth level of life? Before answering yes or no to this question, we must ask three previous questions:

---

[142] *El sello de Eolo*, Edebé, 2000. *The seal of Aeolus*, in press.

1. Has it been proven that altruistic tendencies cannot arise spontaneously by the action of natural selection, without the need for individuals to renounce reproduction?
2. Can there be another way of favoring altruistic over selfish tendencies, without abandoning individual reproduction?
3. Finally, what would a society be like, if human individuals had renounced reproduction, leaving that function, if needed, for a specialized caste?

Before answering the first question, we'll begin by asserting that we cannot apply the experimental method to find an answer, since evolutionary processes are too slow. The appearance of a new species can take a million years. Carrying out large-scale experiments with biological evolution is impossible. We cannot, for instance, take a species of non-social insects (such as solitary bees) and convert it into a new species capable of organizing hives.

Cultural evolution is faster, but even so it would take many years to carry out experiments that, on the other hand, could transgress ethical principles, as they should be performed on human beings. Some experiments of this type have been carried out (for example, in Israeli kibbutzim), although the results haven't always been as expected by the social scientists who designed them.

Fortunately, there is a modern branch of computer science, called *artificial life*, which allows fast experiments of this type by using a technique, genetic algorithms, which simulates biological evolution within the computer. Some of these experiments can be designed to study processes of change in level as those we are analyzing in this chapter, and may shed light on those that took place on Earth over billions of years. In particular, we can get some data to help us answer the question we are considering: *Has it been proven that altruistic tendencies cannot arise spontaneously by the action of natural selection, without the need for individuals to renounce reproduction?*

In an artificial life experiment conducted by me and one of my colleagues[143], we analyzed the question of how evolution can act at two levels simultaneously. The organisms in question resemble ants, but they are not exactly identical. We call them *vants*[144].

Vants live together in *vanthills* and exhibit relatively complex behavior. At birth, they are assigned an expected life span, which can change depending on their activities. Initially there is only one vanthill, located in a two-dimensional territory, through which virtual ants can move. In that territory a number of food sources are scattered, which *vants* can use.

The life cycle of a *vant* is this:

1. It leaves the vanthill and starts looking for food, moving randomly through the territory.

2. If it finds no food, after a while it abandons the search, returns to the vanthill by the shortest path and rests there for some time. Then it goes back to step 1.

3. If it finds food, it takes a part and returns to the vanthill by the shortest path. Upon arrival, it eats a part of the food (which increases the expected duration of its life) and leaves the remainder in the vanthill.

---

[143] M.Alfonseca, J.de Lara: *Two level evolution of foraging agent communities*, BioSystems, Vol. 66:1-2, p. 21-30, June-July 2002.
[144] Contraction of *virtual ants*.

4. If at that time there is at least one other vant in the vanthill, plus a certain amount of food, both can reproduce, resulting in the birth of new vants, who inherit their parents genes, except for the action of two genetic operators: mutation (a random change of some gene) and recombination (shuffling the genes of the two parents).

5. After resting for some time in the vanthill, the vant leaves and returns by the shortest path to the place where it found food the last time. If there is still food there, step 3 is applied again. Otherwise, the virtual ant starts searching randomly, as in step 1.

6. When a food source runs out, it disappears, but a new source appears immediately in a random position.

7. When a vant who knows where there is food (either because it found it in an earlier cycle, or because it is coming back to the vanthill loaded with food) meets another vant from the same vanthill who ignores where there is food, it can act in several ways, controlled, to some extent, by its genes:

    - It can refuse to tell its partner where there is food.
    - It can tell its partner where there is food, but in doing so, it can tell the truth or deceive in varying degrees, sending its partner to the true position, to a position close to the true one, to a random position, or to the diametrically opposite point of the territory.

8. The vant who has received information can believe it or ignore it. This type of behavior is also controlled, to some extent, by the genes of the vant.

9. When the number of vants in a vanthill exceeds a certain limit and the vanthill contains enough food, half of the vants migrate and build a new vanthill somewhere else.

10. When two vants of different vanthills meet in any point of the territory, they can act in several ways:

    - They can ignore each other.

    - If the strongest vant does not carry food and the weakest does, the first vant can take the food from the second. The strength of a virtual ant is a function of its remaining life span, which is subject to genetic control and is also a consequence of the amount of food it has eaten.

    - If the strongest vant does not carry food and the weakest doesn't either, the former can kill the latter and use it as food.

11. When a virtual ant reaches the end of its life span without being able to extend it further by eating, it dies.

12. When the number of vants of a vanthill decreases below five, the vanthill disappears.

It will be seen that the virtual ants of a *vanthill* compete for food at two levels: first, against the *vants* of another vanthill, as the amount of food available at a given time is limited and must be distributed among all. But, secondly, they also compete against those of their own vanthill, for the ant that gets more food will extend the duration of its life and will have more opportunities to reproduce.

Each execution of the previous program is controlled by a number of parameters that receive random values, so two consecutive executions will always be different. This makes it possible to perform a statistical analysis of the results. Figure 11.1 shows an instant during one of the executions of this artificial life program. The letter *a* represents vants that don't know where the food is, and are looking for it. The letter *k* corresponds to vants who know where the food is and are going there. The letter *b* represents vants that are returning to the anthill with food. Finally, the letter *A* represents the position of the vanthill. Note that the vants who know where the food is, and those who are coming back make trails, like real ants.

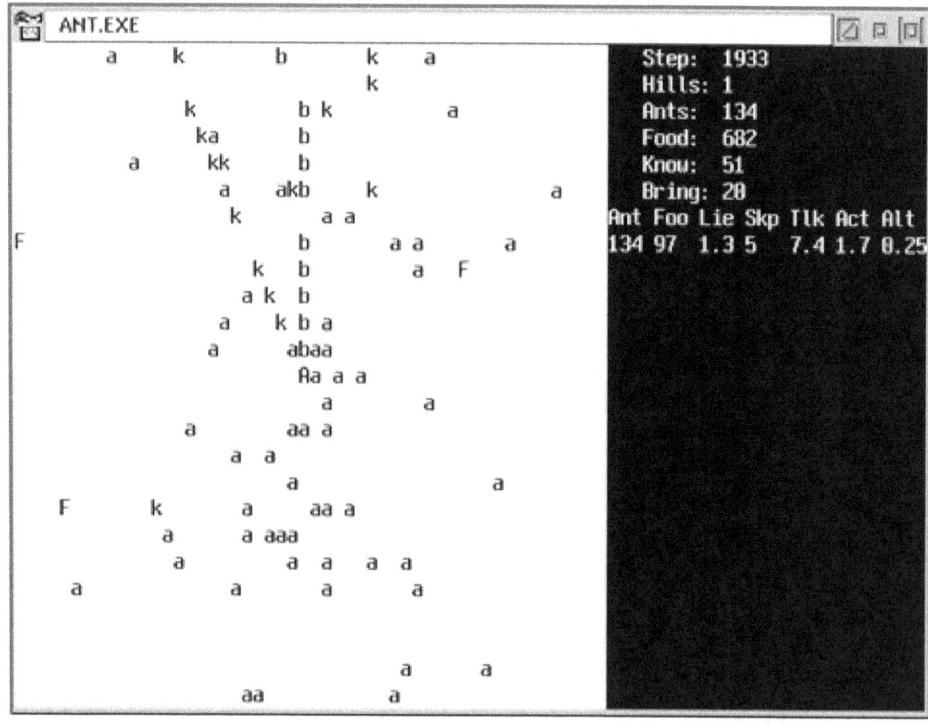

Figure 11.1: An instant in the simulation of virtual ants

As far as we are concerned, the results of this simulation have been spectacular. We have discovered, for instance, that when there is only one anthill, the genetic trait that drives the *vants* to lie to their partners is positively selected, when they would tell the others where they can find the food. It seems obvious why this happens: the ant who lies reserves for itself the position where food can be found, so it will have a better chance of finding it again the next time it visits that place. If it tells its partners where they can find the food, they will share it, and the food will run out sooner, so when it returns it may not find any food there. In this case, therefore, natural selection clearly favors selfish behavior.

On the other hand, when there are several *vanthills* competing with each other in the higher level, natural selection seems to favor altruistic behavior, favoring *vants* whose genes make them tell the truth. It also seems obvious why this happens: if a virtual ant of an anthill insists on selfish behavior, it will extend its own life, but at the expense of that of its companions, and as soon as the number of these descends below five, the whole anthill will disappear.

We see, therefore, that natural selection favors selfish behavior of lower level beings as long as there are not several individuals of the higher level, but otherwise it favors altruistic behavior. Note, in any case, that the *vants* of our experiment have not renounced individual reproduction, which seems to indicate that the answer to the question we asked above should be negative. However, these experiments are not enough to reach that conclusion, because in them we started from the previous existence of both levels, the individual *vant* and the *vanthill*. It's possible that the answer must be affirmative, if we want to study the apparition of the higher level from an environment which contains only individuals of the lower level.

In addition to various scientific conclusions, such as those I have just described, our experiments on artificial life have also provided interesting philosophical arguments. In 2000, I participated as a speaker at the Autonomous University of Madrid in a multidisciplinary seminar-debate on *science and religions: perspectives for a new millennium*. During the debate, another speaker pointed out a typical argument, saying that, after Darwin, we must dispense with the traditional creative God: *Darwin discovered that matter is able to organize itself until it reaches incredible levels of complexity... It seems that the traditional concept of God is no longer valid.*

This reasoning is not necessarily atheist, but if one is not careful, it can be confused with a weak version of another classic argument that denies the existence of God on the grounds that the discovery of evolution makes God unnecessary, for the universe could have come into existence and evolve by itself (!). This argument has a very simple answer, which I used in the debate I have mentioned:

For several decades, a type of computer applications called genetic algorithms has being developed, which use the same mechanisms as biological evolution to produce interesting results. Suppose I make an evolving program that produces artificial intelligence. An intelligent being emerging from my program could use the same argument to prove that I don't exist. As it is evident that I do exist, this shows that the same argument applied to God cannot be valid, for nothing prevents God from creating a universe similar to a large-scale genetic algorithm and having it evolve.

My experiments could also shed light on another classic atheist argument, stated as follows: *How can the existence of a good God be combined with Auschwitz?* It has never come to my mind to intervene actively in my artificial life experiments to modify things as I pleased. It wouldn't be a scientific attitude. It's true that the parallel is not perfect, for I don't think one can say that the universe is an experiment that God makes *to see what happens.* Anyway, I think that this atheist argument denotes an immature attitude, since it does nothing but putting our own blame on God. As Mark Twain said, *there are many scapegoats for our sins, but the most popular is Providence.* In any case, if the Christian vision is correct, God did intervene at least once through the person of Christ, who came to lay our faults upon himself, to become voluntarily our scapegoat.

* * *

Let us turn now to the second question: *can there be another way to favor altruistic over selfish tendencies, which does not imply abandoning individual reproduction?* Even assuming that the answer to the first question may be negative, couldn't the special features of cultural evolution be used to reverse the situation? Couldn't man be an exception to the rule? Haven't we made considerable steps in the path of altruism without renouncing reproduction?

There are several alternatives that could favor altruism against selfishness through various types of social control, if biological control happened to be insufficient. Let's look at a few of them:

1. **Mutual control of individuals**. As we saw in Chapter 8, this procedure was applied by Thomas More in *Utopia*: *Everyone is*

watching you, so you are practically forced to do your work and make proper use of your free time. The trouble is, this kind of control usually works only in small societies, where everyone knows each other. Imagine a modern, overcrowded city, where people who share a home in the same building scarcely do more than greeting one another when they meet by chance, and often don't know what their neighbors do.

2. **Legal and police control.** In the words of the sociologist Pitirim Sorokin[145], every organized group organizes its own meanings, values and norms, which must then be imposed by force. *[The] logical congruence of the norms and actions... of the members is never perfect, but a minimum of them is found in every group, to the extent that it is an organized group. Otherwise, no organized group would be possible.* This type of control, which exists in all our societies, does not guarantee that the goal of norms and values is really altruistic, although it usually is to some extent. However, a specific society can establish regulations aimed at ensuring, not the maximum happiness for all its members, but only for those who belong to a dominant class.

On the other hand, this control procedure gives rise to other problems, such as the fact that human language is always ambiguous, which means that the laws will be subject to different interpretations in different contexts. This can be solved by new laws, which clarify the meaning of the old ones in various circumstances, but the solution gives rise to a new problem: the growth of the legal corpus, which can become so large and complex that few are able to remember all the rules or to

---

[145] Pitirim A. Sorokin, *Society, Culture and Personality*, 1962.

understand them. One of the undesirable effects of this type of control can be expressed as follows: *the same message can have different meaning for the person who gives the order and for the person who executes it*[146].

Strictly legal control structures have the additional disadvantage that they tend to favor bureaucracy and hierarchy, rather than efficiency. The situation does not improve, even when the laws are enacted by democratic procedures, either representatively or directly (by means of a referendum). Regardless of the procedure, there is nothing to prevent the approval of a law that just seeks the benefit of a part of society, rather than that of the whole, even if that part may make the majority of the population.

3. **Control through free market mechanisms**. This would be the typically *capitalist* solution. It assumes that the general interest will always win in the long run, as every member of society will be benefited, even compared to the short term selfish particular profits that can be got by those who just seek their own benefit.

Game theory is a branch of research that has advanced a lot during the twentieth century, although its first achievements date back to the seventeenth. There is a fairly simple, but useful game, called *the prisoner's dilemma*, which can be applied here. In this game, two prisoners, who presumably have been accomplices in the commission of a crime, are offered the following alternative: *If you inform against your partner, you'll be free, while he'll be sentenced to five years, provided that he doesn't inform on you. If this happens, each of you'll be sentenced to four years. Finally, if*

---

[146] Heylighen and Campbell, see note 141.

neither informs on the other, we have enough evidence to sentence you to two years for a less serious crime.

It is clear that, in the long term, the best strategy for both prisoners is not to inform on the other, because this minimizes the time spent in prison. If one of them decides to betray his partner, he runs the risk that the other may also do it, and then both would spend twice as long in jail. But if each of the two prisoners analyzes the situation separately, they'll reach a very different conclusion:

If my partner informs on me, my best option is to inform on him. If I don't, I'll be sentenced to five years; if I do, I'll be sentenced to four. On the other hand, if my partner does not inform on me, my best option is to inform on him. If I do, I am free. If I don't, I'll be sentenced to two years. In both cases, it's convenient for me to betray him. It turns out that the purely selfish and rational analysis of the situation would lead both prisoners to spend four years in jail, when they could have been sentenced to two. Applied to the market, this means that the best solution is not always the one corresponding to a careful analysis of the situation by each participant.

On the other hand, automatic control by market mechanisms assumes that human beings are always perfectly rational. This assumption, however, is not true. We often get carried away by feelings and irrational tendencies. To verify this, it's enough to observe the operation of the Exchange, which in theory is subject only to market mechanisms. Sometimes there are speculative phenomena, where demand rises as the price rises, because less informed investors assume that the trend will continue and risk their assets to get immediate benefits. Paraphrasing a manual of investing tips for the stock market, published several years ago by a Spanish banking entity: *non-professional investors usually hesitate before risking their money, but seeing that the price continues to rise, they finally take the decision to invest, usually just when the price reaches its maximum value. Then, when the price begins to go down, they refuse to sell for fear of the losses, and resist until the price is so low that, desperate, fearing to lose everything, they decide to sell, usually at the time when that the price reaches its minimum value and starts growing again.* Of course, the profits of professional investors on the stock market are nurtured from the losses of non-professional investors who act in this way.

Finally, we know that free market mechanisms can be thwarted through operations such as monopolies or trusts, which allow a few owners to set prices, ignore the general interest and pursue their own. In a book published in 1968, Mancur Olson[147] demonstrates that, in the absence of control, market mechanisms are not enough to ensure the collective benefit that arises from cooperation. His conclusion: there will always be cunning parasites who take advantage of the work of others, so it's necessary to introduce coercive measures to avoid it: what we have called legal and police control.

4. **Validity of a moral law**. This has been the most widely used procedure to ensure the stability of human societies, from the smallest to the largest. Normally it receives the support of a religious belief, and is possibly the most effective of all control procedures, since it favors the predominance of altruistic over

---

[147] *The logic of collective action: public groups and the theory of groups.*

selfish behavior by means of self-control: in the optimal case, the members of a society control themselves.

In the ten commandments of the Law of Moses, a paradigmatic case for our civilization, the altruistic tendency of the moral norms is clear. Although the first three commandments refer to man's relationship with God and can be summed up in a single command (*love God above all things*), the other seven, which refer to our relation with neighbors, command us to obey one's parents, to do justice to others, and to respect life, fame, property and the conjugal life of other people, both in fact and intention. These commandments can also be summed up in a single mandate: *love your neighbor as yourself.*

There's been much talk about the relativity of morals, asserting that each people and time has its own moral code. This conclusion, very exaggerated, has been a consequence of the excessive emphasis placed by anthropologists on sexual morality, which is really just one of the precepts, precisely the most variable.

In a study by the English writer C.S. Lewis, which appears as an appendix in one of his books[148], it is shown that certain basic moral norms are common to all civilizations, all cultures, even those of primitive peoples. Among those rules, the following stand out, which I illustrate with just a small sample of the quotes compiled by Lewis:

- Law of general beneficence: *I have not slain men or ordered others to slain*[149]; *never do to others what you*

---

[148] *The Abolition of Man*, 1947.
[149] Egypt, *Book of the dead*, conjuration CXXV.

wouldn't want them to do to you[150]; *do to others as you wish them to do to you.*

- Law of special beneficence: *natural affection [for the closest relatives] is a right thing and according to Nature*[151]; *I did not use violence with my relatives*[152].

- Duties to parents, elders and ancestors.

- Duties towards children and posterity.

- Law of justice (in court, in ordinary life, etc.): *I did not give false testimony*[153].

- Law of good faith and veracity: *the foundation of justice is good faith*[154]; *You won't betray those who trust you.*

- Law of mercy: *I have given bread to the hungry, water to the thirsty, clothes to the naked, a boat to the boatless*[155].

- Law of magnanimity: *the second [kind of injustice] is failing to protect another from injury when they can*[156]; *death is better... than life with shame*[157].

Lewis's conclusion is that there is a natural law, a moral norm common to all mankind, possibly engraved in our genes, and which we call our conscience. It is not a revolutionary discovery:

---

[150] Confucius, *Analects*, V,11.
[151] Epictetus, I.xi.
[152] *Book of the dead, ibid.*
[153] *Book of the dead, ibid.*
[154] Cicero, *De Off.* I.vii.
[155] *Book of the dead, ibid.*
[156] Cicero, *ibid.*
[157] *Beowulf*, 2890.

this had been the consensus of all mankind until the 19th century, when moral relativism began to spread.

Relativism, whether moral or intellectual, is self-contradictory. Consider the following sentence: *There are no absolute truths*. If this phrase were true, it would deny itself, because it is stated as an absolute truth. Therefore, it must be false. In other words, *there are absolute truths*. As for moral relativism, it has led to many practical contradictions. It is curious that it is precisely during the twentieth century, at the same time as this form of relativism has spread most, when the universal validity of human rights has also been promulgated, in an unconscious exercise of moral absolutism.

But the moral law is not enough, by itself, to ensure the predominance of altruism over selfishness. When someone says *I don't regret anything*, the persons saying this either lie or close their eyes. We must all regret many things, it's inevitable. As soon as we look at ourselves sincerely, we are aware of having failed, not once, but a thousand times, in the strict compliance with our principles, in the perfect adaptation of our behavior to our conscience.

This is the *sense of sin*, which now seems to have disappeared, although only seemingly. Although we don't call it *sin*, we still vilify those who behave in certain ways. Few supporters of moral relativism dare to justify, for instance, racism, genocide, torture, slavery or injustice; therefore they deny their own relativism. As soon as the absolute value of a single principle is accepted, all arguments in favor of relativism lose all their weight.

Remember that we were trying to find an answer to the second question: *can there be another way to favor altruistic over selfish tendencies, which does not imply abandoning individual reproduction?* We have reviewed four possible methods. None of them, by itself, seems able to do it. Perhaps a combination of the four? Some optimistic authors think so. But I doubt it.

* * *

Let's now turn to the third question: *what would a society be like, if human individuals had renounced reproduction, leaving that function, if needed, for a specialized caste?* The answer is expressed quite clearly in one of the most important novels of the twentieth century: *Brave new world*, by Aldous Huxley. In the society described in this novel, sexuality has been totally separated from reproduction, which is carried out exclusively in laboratories, from fertilization until birth. All embryonic development takes place in gestation machines. No one knows their own parents, to the point that the word *mother* has become an obscene word. The education of children is the exclusive concern of the community, the family has disappeared. On the other hand, society has been divided into castes, produced artificially in the gestation machines by administering certain substances to fetuses. Only the upper caste (the *Alpha* type) can donate eggs or sperm to contribute to reproduction. The other castes are not allowed to leave offspring.

In the words of Mr. Foster, one of the characters in the novel: *...in the vast majority of cases, fertility is a nuisance. A fertile ovary for every thousand two hundred would really be enough for our purposes.* The human society painted by this dystopia is very similar to an anthill, an insect society. Life in this society forces its members to total conformity or exile. Is this where we want to go? I suspect not. Is there any other way out? In spite of everything, I'll answer on the affirmative. Which way? We'll consider it in the last chapter, but first we must review certain recent discoveries, some of which could point, if we are not careful, precisely in the direction of *Brave new world.*

12. Can we control our evolution?

As we saw at the end of chapter 6, since man appeared, biological evolution has been progressively replaced by cultural evolution, which is much faster and admits hybridization, feature exchange and dissemination, which take place with difficulty in the strictly biological field. In the last ten thousand years, man's dominance over nature has increased through the control of other biological species (agriculture and livestock), of physical-chemical forces (fire, electricity, magnetism, nuclear energy), of information transmission and processing (communications, computer science) and of our own body (medicine).

During the 19th century, man discovered biological evolution. Just over a century later, we began to be able to direct and take advantage of evolution, by acting directly on one of the basic elements used by natural evolution: DNA (which stores the genetic information). The discovery of the mechanisms that regulate the working of nucleic acids has caused the emergence of new and revolutionary techniques that make possible the genetic manipulation of living beings. These techniques have been given the name *biotechnology* and have led to the emergence of a new industry, genetic engineering, with applications in the fields of pharmacology (drug production), medicine (diagnosis and treatment of genetic diseases) and biology (systematics, taxonomy and decipherment of the genomes of various species of living beings, especially man).

In this chapter we ask whether the application to man of the techniques of biotechnology can lead to a rebound in biological evolution, this time artificial (led by cultural evolution). This could result in an acceleration of the process that is taking us to the fifth level of life.

In order to answer this question, as a previous condition, and to make possible many applications of genetic engineering, it was necessary to know in detail the genome (the genetic composition) of man and other species of living beings. In Chapter 2 I mentioned that the complete sequence of the 5375 nitrogenous bases in the nucleotides of the DNA of a small virus, ΦX174, was obtained in 1976 (this was the first decoded genome). The knowledge of this sequence is important, as we can translate this genetic code and discover which proteins the virus is capable of generating when it takes control of the machinery of the cells it parasites, and what are the mechanisms it uses to reproduce and go from cell to cell.

It's not difficult to imagine that a similar knowledge of the complete sequence of nitrogenous bases in the forty-six chromosomes making up the human genetic endowment would be enormously useful for diagnosing, preventing, and even correcting many inherited diseases, of which we know over four thousand, which affect one in every hundred children being born. Some of these diseases are as important as muscular dystrophy, manic-depressive syndromes, Lou Gehrig's disease, hemophilia, cystic fibrosis, several forms of cancer, hypertension, epilepsy, immunodeficiency and Alzheimer's disease.

The problem, in the case of man, is the tremendous amount of information that must be processed. Each of our cells has forty-six chromosomes, distributed in twenty-three pairs. One of the members of each pair (twenty-three chromosomes in all) is inherited from our father, the other from our mother. Thus, all our cells, with the exception of gametes, the reproductive cells, have double genetic endowment (they are *diploid*). In principle, therefore, it's enough to know the base sequence of twenty-four[158] chromosomes to know the complete human genome. If those twenty-four DNA molecules were stretched (they are wound on a helix) and placed one after the other, they would measure about 2.7 meters.

It is estimated that the twenty-four types of chromosomes contain between 20,000 and 30,000 genes, each of which directs one or more of our biological characteristics. The total number of nucleotides, and therefore of nitrogenous bases (adenine, guanine, cytosine and thymine) is much greater, close to three billion. The sequence of bases, put in writing, would fill two hundred telephone directories or three years of daily newspapers. All this material together is called *the human genome.*

Of those three billion nucleotides, not all correspond to active or potentially activated genes. There are numerous sections on human chromosomes that don't correspond to a gene coding a protein. Some of them are interspersed in the middle of genes and are

---

[158] One of the twenty-three pairs of chromosomes, the one determining sex, appears in the male in two different forms: chromosomes X and Y.

therefore called *introns*. Others make long repetitive series. Apparently, these segments make almost ninety-eight percent of the genome. Until recently, that part was considered unusable, but lately it seems that there can be genes that, instead of proteins, encode ribonucleic acids that act directly inside the cell without being translated. We must also take into account the existence of regulatory DNA, whose function is to control which nearby genes are expressed or inhibited.

The redundancy and apparent uselessness of 98% of the genome is probably as false as the myth that we just use 10% of our brain, which was mistakenly accepted by prestigious personalities and entities such as Albert Einstein and the Carnegie Institution[159].

In 1984, two decades after the decipherment of the genetic code, the complete sequence of some fifty human genes had been obtained, just over two per thousand. Even then it was clear that the task of discovering the complete composition of the human genome was just a matter of time, although considerable effort and expense would be necessary. If we kept up with the previous pace, it would take seven thousand years to complete the work.

Around that time, experts in biotechnology and molecular biology began to talk seriously about the possibility of undertaking a great project. As soon as the project were started, progress would be accelerated, due to the proliferation of dedicated work groups and the development of techniques unknown during the 1980s, which would probably be discovered and allow the deciphering of sequences faster and more reliably. The most optimistic

---

[159] See my article *The myth of progress in the evolution of Science*, http://arantxa.ii.uam.es/~alfonsec/docs/end.htm, 1999.

calculations presumed that the work could be finished by the year 2000.

Enthusiasm was contagious. Walter Gilbert, from Harvard University, 1980 Nobel Prize in Chemistry for having developed a method to find the base sequence in nucleic acids, asserted that the decoding of the human genome is *the biology of the 21st century*.

The size of the project was such that many preparations and discussions were needed before it could get going. On the one hand, it was clear that it would be unfeasible without the participation of the government of the United States of America and, perhaps, of other countries. On the other hand, some scientists involved in other projects looked at it with suspicion, because they feared that the accumulation of resources in one direction would mean less means for their own research. A seemingly trivial problem, but which caused great debate, was the computer language that should be used for the computer support of the project. The enormous amount of data that would be processed made this one of the critical points of the investigation.

In 1987, the United States Congress took the first step by granting seventeen million dollars to the National Institute of Health (NIH), to cover the expenses associated with the implementation of the Human Genome project. The development of tools and methodologies was considered a priority, rather than pin-pointing the genes and deciphering the sequences, which should advance much faster once those tools were available. Thus the project would initially be technological, rather than biological.

In February 1988, the National Academy of Sciences of the United States issued a report advocating the immediate start of the project, giving as a justification, not just the diagnostic and therapeutic benefits that its happy outcome could have for diseases of genetic origin, but also the considerable technological advances that could be foreseen as a byproduct of an enterprise almost as ambitious as putting a man on the Moon in July 1969. To stop adverse criticism, they added that the project should be funded entirely with new budgets, so financing of other research wouldn't be affected. To accelerate its implementation and reduce the political implications of an international negotiation, it would initially be an exclusively project of the United States, although inviting participation of important foreign research laboratories was not excluded.

In 1989, the National Center for Human Genome Research was founded in Bethesda, Maryland, under the auspices of NIH. Its first director (until April 10, 1992) was James Dewey Watson, Nobel Prize in Physiology and Medicine in 1962, along with Francis Crick, for the discovery of the structure of DNA. Watson had distinguished himself among the most active organizers of the project.

Finally, in October 1990, the Human Genome project began officially, with a planned duration of fifteen years and a total budget of three billion dollars. The planned date to have the complete genome sequenced was, therefore, the year 2005. It was a federal initiative, sponsored by the NIH and advised by a panel of external experts, with the participation of many private companies working on biotechnology, including some of the major pharmaceutical and technological firms (Du Pont, Pharmacia, Applied Biosystems), as well as new companies, such as Genentech, Biogen, Genetics Institute, Genomyx, Bios, Genmap, or TransKaryotic Therapies (TKT). The work was carried out in various university centers and research laboratories distributed throughout the United States, some of them newly created, such as the Human Genome Center of California, which was opened in 1989. There also was some collaboration with European and Japanese teams, such as the Center for the Study of Human Polymorphism (CEPH) in Paris. In addition, other countries, such as Germany, Japan, Italy and the United Kingdom, launched their own research programs on the same subject.

Although the final goal of the project was getting the complete sequence of the nitrogenous bases in all human chromosomes, including those segments that don't encode proteins, this was not the initial goal of the project. The nature of the problem made it possible to solve previous goals, useful but simpler, such as drawing a map of the genome. This was the short-term objective adopted for the first five years of the investigation.

To build the map, one mustn't know the complete sequence of bases; it's enough to know in which chromosome each gene is and in what order. This partial information would make it possible to focus research on anomalous genes (those that produce or cause propensity for certain diseases), making some of the practical applications possible with a smaller effort. To get an idea of the difference, we could compare the genome map with a cartographic representation that shows the cities of a country and the streets in every one, while the complete series of base sequences should be compared with the set of addresses (street, number and floor) of all the inhabitants.

The human genome map is not the only interesting genome. A few research projects addressed other living beings. The first success was achieved by the team of Charles R. Cantor[160], who deciphered part of the genome map of the bacterium *Escherichia coli*, whose only chromosome is ten times shorter than the smallest human chromosome. To draw an outline of the map and pin-point each component, the chromosome was divided into twenty-three sections. Work also started early on the mouse genome, because this animal is frequently used in laboratory experiments to investigate genetic diseases. The results are then extrapolated to man.

The technique used to map chromosomes tried to find reference points, called *markers* or *Sequential Tag Sites (STS)*, in each chromosome. This was the procedure: the chromosome is split into small pieces by means of enzymes called *restrictases* or *restriction endonucleases*, which cut the DNA strands at specific sequences, and a base sequence is found in each segment that does not appear in any other fragment in the same chromosome (a *marker*). If the chromosome is cut successively at different places, it is possible to deduce the order in which the markers are found in the chromosome.

Once the markers have been found, we can find the position of a specific gene that produces an inherited disease. To do this, we take two persons from the same family, intimately related (parents and children, siblings, etc.), one of whom has the disease, while the other is healthy. Their chromosomes are then compared to find differences: the two complementary chains of each chromosome are separated and one of the chains of one person is allowed to join with the complementary chain of the other. The identical areas will

---

[160] Then at Columbia University, New York.

adapt perfectly, while those that differ cannot be associated properly, so the chromosome will be irregular at that point. It is enough then to pin-point the pair of markers enclosing the difference, to discover in which chromosome is located the gene of the disease, as well as the section of the chromosome. To locate the markers, short segments of DNA are used, complementary to them, which adhere to the chromosome just where the marker is located. These segments are called oligonucleotide *probes*, and are usually radioactive, so that their presence can be detected easily. Probes are manufactured today automatically, in large quantities, by means of gene machines.

The procedure seems simple, but in fact it is harder, for there can be many differences between two persons of the same family, in addition to the one causing the genetic disease. The experience must be repeated with other sick-healthy couples, to eliminate the differences gradually until the culprit is found. In addition, for the procedure to be successful, the markers must be well chosen, as they must appear in the genetic endowment of both persons.

An additional complication of the problem is genetic recombination. When cell division occurs, sometimes a chromosome is split into two or more pieces and rebuilt again, but the pieces are not always joined in the same order. This may mean that the markers, which in one individual are in a certain place, can be in a different position in another person. It may also happen that, at the time of recombination, the paternal and maternal components of the chromosomes exchange segments, shuffling the genetic information. Consequently, the search for a gene is a very slow process and affects dozens of markers, which must be carefully studied one by one.

Thus, the first step to build the human genetic map was localizing a large number of stable markers for each chromosome. The search would be easier if the markers were regularly spaced. The distance between two markers (or between two genes) of the same chromosome is measured in Morgans[161], equivalent to one hundred million nucleotides. The initial objective of the *Human Genome project* was to find a set of genetic markers separated by distances not exceeding one centiMorgan (1 cM), so that there were no more than one million nucleotides between two consecutive markers. Since there are about three billion nucleotides in the human genome as a whole, the number of markers needed was close to three thousand. At the beginning of the project, about four hundred were known, equivalent to an average distance between markers of 8 cM. In 1993, the Center for the Study of Human Polymorphism (CEPH) in Paris, in collaboration with the Pasteur Institute and other research entities, announced that they had built a map with two thousand genetic markers, equivalent to a distance between markers of 1.5 cM.

In 1983, the American Kary B. Mullis discovered the polymerase chain reaction (PCR), a technique that makes it possible to get as many copies as desired of a given DNA fragment. This discovery, which earned him the 1993 Nobel Prize in Chemistry, has made possible DNA analysis, which has revolutionized medicine, justice and criminal investigation techniques. The reaction is simple, based on an enzyme (polymerase), which the cellular machinery normally uses to repair damaged DNA. This enzyme can lengthen

---

[161] In honor of Thomas Hunt Morgan (1866-1945), 1933 Nobel Prize in Physiology and Medicine, considered the father of modern Genetics because of his experiments performed mainly on the fruit fly (*Drosophila melanogaster*).

a DNA fragment, if it is associated with its longer complementary chain. Thus, if one of the two branches of a chromosome looses a few nucleotides, but the other branch remains unchanged, the polymerase makes the damaged piece grow, precisely with the succession of adequate bases to return to normal.

The polymerase chain reaction takes advantage of the normal operation of this enzyme: one or more copies of a double molecule of DNA, one of whose fragments must be reproduced, is introduced into a test tube. Polymerase is added, along with many copies of certain oligonucleotides (tiny DNA molecules) called *primers*, that correspond to the ends of the fragment that one wants to reproduce. By heating this mixture to just under 100°C, the complementary strands of DNA are separated. Then the temperature is lowered to about 60°C, which lets the chains pair again. Since there are many primers, they are coupled with the longer chains. The polymerase action extends them, thus reproducing the fragment to be copied. The mixture is then reheated and cooled again, and the process is repeated, starting now from twice as many copies of the desired fragment. Repeating the cycle over and over again, at each stage the number of copies of the fragment is doubled, which results in an exponential growth that, in a few hours, can multiply their number by one hundred thousand.

In 1991, the team of J. Craig Venter, who then worked at the *National Institute of Neurological Disorders and Stroke* of Bethesda (Maryland), invented a technique that uses messenger ribonucleic acid (RNA) to identify segments of genes, which then serve as markers in chromosomal DNA, thus pin-pointing them quickly. The use of messenger RNA ensures that the fragments correspond to genes, not to filler segments that don't encode proteins. In addition, we know that those genes are active at the time of being pin-pointed. Venter worked with brain cells, which use a large number of different genes, many of which are not used in the rest of the body. His works made it possible to find markers for about six hundred different genes, many of which were already known, although two hundred were new. This technique accelerated the localization of genes by several orders of magnitude, proportionally reducing the cost.

In 1987, Maynard V. Olson and David Burke discovered that it was possible to insert in yeasts artificial chromosomes (YAC) made of synthetic DNA, which worked like normal yeast chromosomes and reproduced with them. To do this, a YAC contains certain very short sequences extracted from the normal DNA of the yeast, which makes it possible to deceive their cellular machinery. The rest of the artificial chromosome can be as large as desired, at least in theory. In particular, one can introduce there a human gene, or a part thereof. The YAC were used successfully to accelerate the search for the human genetic map, although they were later replaced by BAC (*Bacterial Artificial Chromosome*) based on bacteria, with better results.

Using this method, two teams of researchers managed in 1992 to obtain the almost complete map of the two smaller human chromosomes: Y and 21, the latter especially interesting because it contains the genes causing amyotrophic lateral sclerosis (Lou Gehrig's disease), progressive myoclonic epilepsy, and a form of Alzheimer's disease. In addition, Down syndrome (Mongolism) is a consequence of the presence of a supernumerary copy of this chromosome in the cells of an individual.

In 1998, a new private company challenged the Human Genome project by launching a parallel and independent research project, aimed at the same objective. The company, Celera Genomics, had been founded by J. Craig Venter, who had quit his job with the NIH. To carry out his plans, Venter devised a procedure to obtain base sequences, called *shotgun*, with the following steps:

1. The process starts by selecting several copies of the set of chromosomes (genome) to be analyzed.

2. Chromosomes are broken into small pieces by means of restriction endonucleases, as explained above. Different copies of the chromosomes are broken at different places, so the pieces won't be identical.

3. The base sequence of each segment is obtained using automatic machines.

4. A computer program compares the sequences, looking for common areas. This allows the pieces to be sorted and the base sequence of the entire genome is got.

Since 1986 there are automatic machines that extract the base sequences in DNA fragments. The first was designed and produced by Applied Biosystems. The model used by Celera had just been developed by Perkin-Elmer, which became a part of the new company. Collaborating with Hamilton Othanel Smith[162], and using the shotgun procedure, by 1995 Venter deciphered the first complete genome of a second-level living being: the bacteria *Haemophylus influenzae*. A few months later, using the same

---

[162] 1978 Nobel Prize in Physiology and Medicine, having found a new type of restrictases.

procedure, he obtained the genome of a simpler bacterium, *Mycoplasma genitalium*.

The shotgun procedure accelerated the deciphering of genomes of living beings (see table 12.1). Since 1995, milestones were reached always faster. 1996 saw the decipherment of the genomes of yeast (a single-celled eukaryote, therefore a living being of the third level), and of the archaea *Methanococcus jannaschii*. In this way genomes of the three large groups of living beings were then known: archaea and bacteria (prokaryotes) and eukaryotes. In 1997 the genome of *Escherichia coli* was published. In 1998, the first animal, the nematode *Caenorhabditis elegans*. In 2000, a partial version of the genome of *Drosophila melanogaster*, the famous fruit or vinegar fly, used in inheritance experiments. That same year, the first plant, *Arabidopsis thaliana*, was deciphered. In 2001, the first fish, *Fugu rubripes*. In 2002 the mouse, and in 2004 the rat.

Year	Species	Type	Nr. bases	Genes	Chromosomes
1976	ϕX174	virus	5375		
1995	Haemophilus influenzae	bacteria	1,830,121	1,749	1
1995	Mycoplasma genitalium	bacteria	500,000	470	1
1996	Saccharomices cerevisiae	yeast	12,500,000	6,000	16
1996	Molluscum contagiosum vir.	virus		163	
1996	Methanococcus jannaschii	archaea			
1997	Escherichia coli	bacteria	4,638,858	4,300	
1998	Chlamydia trachomatis	bacteria		900	
1998	Caenorhabditis elegans	nematode	97,000,000	20,000	

2000	Neisseria meningitidis	bacteria		2,100	
2000	Drosophila melanogaster	insect	180,000,000	14,000	
2000	Human (first outline)		~3,000,000,000	35,000?	46
2000	Vibrio cholera	bacteria		3,885	
2000	Mycobacterium tuberculosum	bacteria		4,000	
2000	Arabidopsis thaliana	plant	125,000,000	27,000	
2001	Human (map)				
2001	Staphylococcus aureus	bacteria			
2001	Fugu rubripes	fish	400,000,000	30,000	
2001	Yersinia pestis	bacteria			
2002	Rice	plant		30,000-55,000	
2002	Mouse	mammal	2,500,000,000		40
2002	Anopheles gambiae	insect		14,000	
2002	Plasmodium falciparum	protozoan		5,300	
2003	Human (error 1/100,000)		~3,000,000,000	35,000?	46
2003	Dog (partial)	mammal			
2004	Rat	mammal			
2004	Poplar	plant	480,000,000		19

Table 12.1. History of genome deciphering

Meanwhile, the competition between the governmental project and the private company, together with the application of new methods, accelerated the deciphering of the human genome, which exceeded the most optimistic forecasts. In 1998, the directors of project advanced their planned completion to 2003 (instead of the initial 2005), announcing that by 2001 a complete genome outline would be available. This outline was published in 2000, having been obtained practically at the same time by the two participants in the race. Work continued, however, and in 2003 a new, more complete and accurate version was published, with a maximum of 30,000 errors: on average, one erroneous base per 100,000 nucleotides (one milliMorgan).

One of the key problems of the *Human Genome project* was computer processing. On the one hand, it was necessary to build databases of the sequences, to detect whether those newly found were previously known genes or additional information. Sequence search was not a trivial problem. The number of characters in the entire genome (if each base is represented by one character) is close to three billion, equivalent to three gigabytes of information. In 1986, when the project began to be discussed, it took a personal computer (IBM PC) one hour to locate a particular sequence among those contained in a database with four million nucleotides. In 1993, the speed of personal computers had multiplied by fifteen, but the number of available data had increased by the same or greater proportion. Only at the end of the decade very fast computers appeared, capable of storing gigabytes of information.

To speed up the process, fast sequence comparison methods were found. In particular, a technique invented by IBM researchers in the United States for spell checking dictionaries associated with word processors, was also used in the *Human Genome project*. This method, known as FLASH[163], makes it possible to locate quickly a particular sequence, together with similar ones. In 1988, the author of this book suggested to Dr. Julian Davies of the

---

[163] Initials of *Fast-Lookup Algorithm for Sequence Homology*.

Pasteur Institute in Paris that this technique could be used to find genetic sequences.

Computerized databases can be used, not just to identify new sequences; they also make easier the generation of gene copies, which can be used in pharmaceutical and clinical analysis. Initially, known gene models were stored as DNA molecules in liquid nitrogen, but as the number of deciphered genes increased, this form of storage proved prohibitively expensive. At present the databases are connected to a DNA synthesizer, which automatically generates the selected sequences to be used in experiments.

Once the human genome has been sequenced, what comes next? When the genome of a living species is known, what can be done with it? Several things:

- For a pathogenic microorganism, knowledge of its genome makes it possible to make vaccines or medicines directed exclusively against that organism or against any of the proteins it uses, which wouldn't cause side effects to the patient. This is the reason why, among the genomes obtained first, there are those of many pathogenic organisms, such as bacteria that cause infections (*Haemophilus*, *Staphylococcus*), venereal diseases (*Chlamydia*), meningitis (*Neisseria*), intoxications (*Salmonella*) and those organisms that caused terrible epidemics in the past (and a few at present), such as cholera (*Vibrio*), plague (*Yersinia*), tuberculosis (*Mycobacterium*) and malaria (*Plasmodium*).

Some of these new vaccines have already been obtained by inserting genes of the corresponding pathogenic organisms into the *Vaccinia virus*, which is the base of the smallpox vaccine and was used by the English doctor Edward Jenner (1749-1823) for the first vaccine in history, which two centuries later eradicated forever human smallpox throughout the world. This success prompted researchers to imagine that the same virus, genetically modified, could be used to combat other diseases. Among those obtained by this procedure, the hepatitis B vaccine was the first to reach the market. There has been later work on new vaccines against influenza, malaria, rabies (hydrophobia), and herpes simplex. Polyvalent vaccines can also be made, combining in a single virus antigens against different microorganisms: tests have been done with a joint vaccine against hepatitis B, herpes simplex and influenza.

- A gene belonging to a different species (such as man) can be introduced into an easily manageable living being in the laboratory (such as *Escherichia coli*) and the bacteria are then induced to produce the corresponding protein in large quantities. The gene is introduced in the bacterial genome by mechanical or biological procedures, using vectors belonging to the first level of life (viruses or plasmids). Since the 1980s, these procedures are the basis of biotechnology, and have made it possible to synthesize large amounts of substances and medications previously scarce or difficult to obtain, such as human insulin, growth hormone, erythropoietin, interferon, interleukins, tissue plasminogen activator (an enzyme essential for the dissolution of blood clots), protein C (a human anticoagulant used in the treatment of cardiovascular and coagulation problems), tetracycline, the factor of tumor necrosis (which is used against cancerous tumors), blood coagulation factor VIII (the lack of which causes hemophilia type A), and many others.

- Using the same procedures, strains of organisms capable of performing various industrial works can be obtained. In 1981, the bacterium *Pseudomonas putrida* was genetically manipulated to improve its ability to degrade oil. In 1983, a gene originally belonging to *Thermomonospora* bacteria, which allows it to digest cellulose, was transplanted to *Escherichia coli* for use in the decomposition of paper waste. That same year it was possible to combine the genes of two different bacteria that direct the production of indigo, which had never been produced synthetically in practice. Neither of the two bacteria was able to generate it, but the recombined bacteria did.

- The genome of mammals used in the laboratory (rat, mouse) can be compared with humans, to estimate whether the results of experiments on genetic ailments performed with one species can be applied to the other.

- For organisms of economic interest (such as rice or cows) and for man himself, knowledge of the genome can let us find better treatments for hereditary ailments. The correction of genetic defects (genetic therapy) could cause a revolution in medicine similar to those caused by the discovery of anesthetics in the nineteenth century and antibiotics in the twentieth. Many genetic defects are due to the absence of a normal gene in the DNA of the cells of a living being. If it were possible to reinsert the missing gene and get it properly expressed at the indicated place and time, the defect would have been corrected.

- New proteins can be designed to obtain favorable effects, by introducing arbitrary mutations in the genes that encode them and subjecting the resulting proteins to accelerated evolution in the test tube, by selecting those among them that produce the desired effects. This procedure, which is called directed molecular evolution, similar to what we do in the computer in artificial life experiments, has been used to find better additives for detergents, which prevent dyed garments from staining other fabrics subjected to the same washing process.

- Finally, although for the time being this is science fiction, it may be possible someday to improve living species (man included) by causing controlled mutations in embryos (changing one or more bases at certain points of the genome) or by introducing genes belonging to one species into another (this has already been done). In this way, cultural evolution would lead to biological evolution through processes equivalent to hybridization and the exchange of characteristics between different species, freeing this type of evolution from its inherent limitations and raising it, in a way, to its own level.

- A non-trivial problem is the definition of what is meant by the human genome. The genome of a particular individual?

Which one? For the Celera project, Craig Venter decided to use his own DNA. But who is the middle man? We know that there are differences between the genetic endowments of any two individuals. It will surely be necessary to obtain genomes from many people, so as to compare them with each other.

By knowing the sequence of the bases we are still far from having reached our goals. First of all, we must pin-point each and every gene. We know where many of them are, but not all. Then we must find out what each gene does, considering that a gene can take part in many vital processes, not just one. There re many things to discover: we are far from being able to direct our own evolution.

For instance, we have no idea what should be modified to increase the altruism of human beings, because we don't understand the relationship of our behavior with our genes, if there is any. Therefore, trying to favor the appearance of the fifth level of life by manipulating our genes is a very far away objective, assuming it is possible. However, at the speed with which biotechnology advances, we may be there sooner than we think. In that case, we should ask ourselves a couple of questions related to ethical issues that may arise in this regard. We'll talk about this in the next two chapters.

13. Should we control our evolution?

We have talked in the previous chapter about the possibility of genetic engineering leading us, before long, to design controlled changes in embryos. These changes, conveniently chained, would make possible to direct the biological evolution of the human species along predetermined paths, rather than leaving it to chance. For the first time in the history of life on Earth, a biological species would be able to guide its own future destiny. This raises two key questions. In this chapter we'll address the first: *if it were possible to reach the fifth level of life through genetic manipulations, should we do it or, on the contrary, would it be better to stop it for ethical reasons?* In the next chapter we'll move on to the second question, also very important: *does the fifth level imply an ethic?*

Let's start with the first question, already raised in antiquity, when Plato proposed in *The Republic* that the same methods used since time immemorial for the selection of domestic animal breeds could be used to improve human beings. But it was the English scientist Francis Galton (1822-1911), cousin of Charles Darwin, who in 1869 coined the term *eugenics*[164] for an ambitious program that envisaged the selection of matching people, who should be encouraged to have offspring, so as to take us along a path of directed evolution that would improve intelligence and physical conditions.

---

[164] From the Greek *eu*, good, *genea*, birth, offspring, lineage, origin; with the meaning of *good inheritance*.

During the twentieth century, the principles of eugenics tempted personalities such as George Bernard Shaw and were studied, with dubious scientific methods, by the Eugenics Record Office (Cold Spring Harbor Laboratory, United States, 1910) and adopted as a political program by associations such as the *American Eugenics Society*, founded in 1926, which held that the upper classes are actually superior, because their members enjoy a better genetic endowment. All these activities led, in the first decades of the twentieth century, to thirty US states passing mandatory sterilization laws, applicable to criminals, subnormal and *degenerate* such as epileptic, blind, deaf, deformed and others. It is estimated that the application of these laws led to the sterilization of some sixty thousand people.

Similar laws were passed and applied in Switzerland and some Nordic countries, but they found an optimal breeding ground in Hitler's Germany, which quickly passed from mandatory sterilization[165] to euthanasia, and used eugenics as an argument for the extermination of homosexuals, gypsies, blacks and Jews. Consequently, eugenics was discredited after the Second World War, although it resurfaced in the last decades of the twentieth century, when the laws of many countries became permissive with abortion, accepting as justification the detection of possible genetic defects in embryos. Eugenics and abortion have become flags of the political left.

* * *

---

[165] The Nazi law, approved in 1933, gave rise to the sterilization of several hundred thousand people.

The Fifth Level of Evolution

Science is never neutral, from an ethical point of view. It's true that objects or discoveries are neither good nor evil. Only human acts can be good or evil. But tools, scientific discoveries, tend to increase the power of man over the environment. They can be used well, but they can also be abused, for they are just instruments. A hammer, for instance, can be used to place a work of art where everyone can see it, or it can be used to destroy it. An atomic bomb could be used to erase a city, or to divert an asteroid that threatens to crash into Earth, saving millions of lives.

The same applies to genetic engineering: it can be used as an instrument to obtain beneficial effects, such as the correction of hereditary diseases or cancer, a most effective production of food, or designing cheap medical drugs; but it can also put man in very dangerous situations. In this chapter we'll review a few of them.

Certain genetic manipulations seem morally acceptable. Consider the possibility of correcting inherited diseases by inserting in the patient a missing gene. The question is quite different when we speak about other types of manipulation. Except in a few aberrant cases, medical science has always opposed experiments on human beings, except in very specific situations and under strictly voluntary conditions. Now we can see a new type of eugenics emerging on the scientific horizon, based on embryo manipulation, so as to direct the biological evolution of the human species by making non-therapeutic changes in its genetic composition. This form of experimentation cannot be voluntary, the changes would be irreversible, and the subjects will never be able to give their opinion.

Who would control this so-called *self-controlled evolution*[166]? Human beings? But who are *human beings* in this context? The phrase *man will control his evolution* is equivocal or false, if taken at face value. What would actually happen is that a few would control the evolution of others. Once the problem is expressed in these, its true terms, everything is clearer: nobody should have the right to modify the biological constitution of other people. Genetic manipulation would become, in this case, an extreme version of brainwashing, potentially even more dangerous.

All human generations have tried, to some extent, to control the next generation. Education is a powerful weapon, but fortunately incomplete, since human freedom allows individuals to rebel against the way they have been educated. Genetic manipulation is a much more subtle form, which perhaps will also fail, but which can provide a greater, fearsome control. Remember, once again, the novel *Brave New World* by Aldous Huxley.

* * *

Since its inception, a few decades ago, genetic engineering has posed ethical problems, possibly in greater proportion than any other branch of technology, present or past. In the following pages we will analyze a few of them, so that they serve as a sample of the complex problems that may arise from embryonic eugenics.

The first example considers the possibility that the genetic manipulation of a non-human organism causes harmful effects on humanity. Let's look at one of the simplest cases: suppose that a certain experiment inserts, perhaps inadvertently, the gene of a

---

[166] Juvenal: *Sed quis custodiet ipsos custodes?* But who watches the watchers?

very potent toxin in *Escherichia coli*, a bacterium frequently used in the laboratory. Suppose some of these manipulated bacteria escape from the controlled environment where they were created and spread among humans. Recall that *Escherichia coli* lives naturally in our intestines in a state of symbiosis (common life with mutual benefits), although sometimes it becomes pathogenic. The modified bacteria could have some advantage over their normal congeners, for instance, resistance to an antibiotic. The spread of an artificial bacterium with these characteristics could lead to an epidemic catastrophe unprecedented in history.

Some critics say that researchers must have gone crazy by choosing precisely *Escherichia coli* to conduct biotechnology experiments. Others, however, claim that this is safer, for we know much more about this bacterium than about any other microorganism. It is argued that the K-12 strains of *Escherichia coli*, which are used in the laboratory, are no longer able to live in the human intestine. New strains have also been created, with significant genetic deficiencies, which don't let them survive outside the controlled conditions in which they are maintained. The argument has weight, but it's not final, as these incomplete strains could recover their ability to survive by recombining their genetic composition with normal bacteria, which often occurs naturally.

Anyway, given the possible magnitude of the problem, in 1974 some researchers made the voluntary decision to temporarily postpone several potentially risky tests. One of them, especially important, was using the *shotgun*[167] method to divide the whole DNA of a living being by means of restriction endonucleases. The

---

[167] See chapter 12.

fragments would then be separated and recombined with a vector (a plasmid or a virus), which would be introduced into a specific *Escherichia coli* cell, which would then be cloned separately, obtaining from each a different colony. The complete genetic endowment of the initial living being would be distributed among all the colonies, making it possible to experiment with it, cross various strains and perform all kinds of manipulations.

In 1975, scientists and researchers from the then emerging field of genetic engineering met in Asilomar to discuss these issues. The results of the conference were used by NIH (National Institute of Health) to develop standards to classify experiments, separating the safest from the most dangerous (which were banned), through a progressive scale of risk levels and a corresponding increase in physical and biological protection measures. The risk levels were the following:

- **P1: minimum risk**. They use DNA from non-pathogenic organisms (plasmids or common bacteriophage viruses), which recombine spontaneously with *Escherichia coli*.

- **P2: small risk**. They use DNA from embryonic cells belonging to cold-blooded vertebrates and lower eukaryotes (except insects); plants (except those that produce pathogens or toxins); or low-risk pathogenic prokaryotes that exchange genes spontaneously with *Escherichia coli*.

- **P3: moderate risk**. They use DNA from plant viruses or non-pathogenic prokaryotes that do not recombine spontaneously with *Escherichia coli*. Also included in this group (although greater security measures are required) is DNA from embryonic primate tissue, mammalian and bird cells, toxic vertebrates, or animal viruses (when the recombined DNA does not contain harmful genes).

- **P4: high risk**. They use DNA from non-embryonic primate cells (because of their proximity to man), or from animal viruses (when the recombined DNA does contain harmful genes).
- **Totally forbidden**. The experiment mentioned using the *shotgun* method; release of manipulated organisms into the environment.

Like all rules, those of the NIH left no one satisfied. Many researchers considered them exaggeratedly restrictive, while for others they were too lax. Of course, the dangers due to a possible misuse of biotechnology are very serious, and can affect not just man (through hypothetical diseases and epidemics), but to the entire biosphere (by causing ecological imbalances). There is also the danger that these techniques can lead to the production of biological mass destruction weapons, which could fall in the hands of terrorist organizations.

During a public session organized in March 1977 by the Washington Academy of Sciences, some of the participants said that the issue should not be left to scientists, but to politicians. According to these critics, research should be centralized and the most dangerous would be carried out in carefully chosen places with strong security measures. Many researchers declared against this, as they'd be isolated from other scientists and the cost of the experiments would increase. They argue that explosives are also dangerous and prone to misuse, but their use is not forbidden, just regulated.

As a result of the debate, some U.S. states, the Senate and the House of Representatives discussed possible laws to regulate biotechnological experimentation. The legislation must be provisional, for knowledge about possible risks would increase as experimentation progressed, and legislation must be refined or corrected in one or another way.

For a decade, things remained like this, although certain citizen movements systematically opposed by legal means all the steps being taken in the direction of greater liberalization of research. These lawsuits were able to delay the tests for several years in some cases. Be that as it may, the standards gradually relaxed, and on April 24, 1987, the first release into the environment of a genetically engineered organism was carried out: the bacterium *Pseudomonas syringae*, which normally lives in the soil and on plants and is responsible for frost on cold mornings, for ice cannot form unless it finds a crystallization nucleus, which is provided by this bacterium and similar species.

Frost has harmful effects on crops and causes losses of one billion dollars a year in the United States. The presence of nucleating bacteria could be fought with streptomycin, but spraying plants with antibiotics is not convenient, as its effect could be even more adverse than frost. For this reason, the company Frost Technology tried to apply biological control without genetic manipulation, placing on the market a bacteriophage virus that infects only these bacteria. In parallel, the University of California in Berkeley, in collaboration with Advanced Genetic Sciences of Auckland (California), genetically manipulated *Pseudomonas syringae*, eliminating the two genes that provide it with the ability to act as ice condensation core.

In fact, nobody knew which enzymes were responsible, but the genes directing their formation were discovered by chopping the bacterial chromosome and inserting the pieces one by one in *Escherichia coli*, until a colony was found that could produce ice. Then that DNA fragment was removed from the chromosome of *Pseudomonas syringae* giving rise to a version of this bacterium that doesn't generate frost. The altered bacteria, almost identical to the natural ones, would be sprayed on the crops to be protected, just before their wild congeners start their annual cycle.

Although it was a low-risk case (because nothing had been added to the bacteria, just something had been removed), it took more than a year for the NIH to approve the first controlled outdoor test. There was some discussion about possible risks of climate change, as this species of bacteria also lives in the atmosphere, carried away by air currents, and is involved in the formation of ice crystals (snow). However, this was unlikely, for mutilated cells couldn't have genetic advantages that would let them supplant their normal competitors in large areas or around the world, while the experiment would be carried out in a very small area. In addition, certain tests carried out by Monsanto with an altered strain of *Pseudomonas*, (two genes without special function were inserted, just to make them easy to find), discovered that bacteria barely move from where they are sprayed (no more than 35 centimeters).

Although the approval was granted in 1983, the test had to be postponed several times due to successive lawsuits and appeals, which managed to delay it until 1987. On that date a crop of altered *Pseudomonas syringae* was sprayed on open-air strawberry trees. The experiment was a success: the bacteria did not spread outside the assigned area and the freezing temperature dropped by two degrees.

In October 2004, the success of an experiment classified at the P4 risk level was announced: the artificial reconstruction of the virus that caused the deadly influenza epidemic in 1918. Other more dangerous experiments have been banned. In February 1994, the British government paralyzed a research of the University of Birmingham about the insertion of oncogenes (genes that increase the propensity to cancer) in adenovirus (which causes the common cold), so as to introduce them in cell cultures and study their effect. Obviously, if the altered adenoviruses would escape from the laboratory, the risk would be large. The measure seems reasonable, despite the protest of the scientists, who argued that their security controls were sufficient and effective.

* * *

Another important ethical problem related to genetic engineering is the granting of patents for living beings designed by researchers. Since the techniques of genetic recombination were developed, companies in the field have tried to patent various strains of bacteria, to ensure exclusive rights in their exploitation. However, the US patent administration had reasonable doubts about the desirability of granting a patent on a living being, since the legislation had been designed to protect inventions, normally built with inanimate materials, and it was not clear how it could be applied to viruses, bacteria, plants or animals. These delays induced biotechnology companies to resort to industrial secrecy to protect their activities.

The issue is especially difficult to resolve, since the boundaries between what may or may not be patented are not well established. It seems clear that a gene that exists in nature should not be patented, as a mineral deposit or a new star cannot be patented, since its location cannot be considered an invention, but a discovery. On the other hand, it could be argued that a synthetic gene, totally new and non-existent on Earth, does meet the conditions required by patent laws.

Once the possibility of patenting synthetic genes was accepted, the extension of the patent to living beings that contain them was proposed, based on the fact that they do not exist in nature. However, this is also debatable: living beings change continually their genetic endowments as a result of spontaneous mutations, and no one had thought of protecting legally their use and exploitation. A similar problem would arise in the case of hybrids and plant races designed to improve crops. In this field, a consensus similar to the rules for the protection of intellectual property was reached in 1970.

In practice, different patent organizations started to accept specific cases, thus establishing historical precedents. The first clearest cases did not describe living beings, but the techniques used for their modification. This is how American researchers Stanley N. Cohen and Herbert W. Boyer patented a method they had discovered to insert genes into bacteria by manipulating plasmids. Subsequently, a patent was granted to the manipulation of *oleophage* bacteria[168], as well as several breeds of plants modified by genetic manipulation, such as some forms of potatoes, cotton

---

[168] *Pseudomonas putrida*, which degrades petroleum.

and tobacco, which are able to resist certain viruses, herbicides and insects.

In April 1988, the United States Patent Office set an important precedent by granting legal protection to a breed of mice obtained by genetic manipulation. A human gene responsible for the propensity to certain forms of cancer had been inserted into the ovules of their mothers. The Harvard mouse, as it was called, would be useful for cancer research in the laboratory. This case opened the way for patents to be accepted for all kinds of non-human living beings. This can be advantageous, but also inconvenient, and may lead to abuse. In fact, the patent for the Harvard mouse, held by the pharmaceutical multinational Du Pont, caused delays and impediments in cancer research, due to the conditions imposed by said company to authorize the use of that breed of animals in the laboratory.

The next phase of the problem, perhaps the thorniest, is deciding whether patents should be banned for human genes. In 1992, C. Thomas Caskey, of the Houston School of Medicine, detected the gene responsible for the formation of brain convolutions in man and tried to patent it. At the same time, the NIH submitted three applications to obtain the patent for a total of 6122 human genes. The debate went beyond the patent office and led to the creation of a commission in the United States Congress. In the opinion of many, the solution is obvious: human beings must not be patentable; no one can claim special rights over other people or any part of their body, including genes. However, the issue is quite difficult and discussions will continue, no doubt, for a long time.

* * *

Another important issue concerns the use of information about the genome of a concrete person by external entities. For example, a company might require that the candidates to occupy a job be tested to determine if they have a genetic defect, and make a negative result a condition for their hiring. This scenario is presented and denounced in the science fiction film *GATTACA*[169], where the action takes place in a not well defined future, when genetic engineering has advanced to such an extent that a drop of blood or a person's hair is enough to get, in a matter of seconds, all their genetic data, their identity and their professional classification, enabling people to perform certain activities and forbidding them from others. Naturally, the protagonist manages to deceive the system and becomes an astronaut, a profession for which a perfect genetic composition is demanded, in spite of the fact that his own genetic composition had relegated him to the role of cleaning employee, and predicted for him a maximum lifespan of thirty years, which of course he manages to overcome.

GATTACA's predictions are not as far away as they may seem. The tests required by companies from their potential employees are ever more complicated, especially in the United States, where some have even been subjected to the lie detector. There are rules forbidding companies from rejecting a job application because of the possibility of suffering a future disease, but these rules have ill-defined legal exceptions. In fact, several flagrant cases have attracted the attention of sociological researchers, who predict that genetic discrimination is imminent.

---

[169] The title of this film is a DNA string with seven nucleotides.

Advances in genetic engineering are spectacularly rapid. At the end of 2003, the possibility was announced of simultaneously detecting thousands of anomalous genes in the genetic constitution of a person, by means of sensors made of silicon wires, 20 nanometers[170] in diameter, to which a DNA molecule designed to detect a concrete mutation can be attached. By mixing a large number of sensors of this type, it would be feasible to detect many mutations simultaneously.

In a few months, these techniques had advanced so much that in May 2004 researchers from the Motorola company announced the design of a miniature laboratory, little larger than a credit card, which detects abnormal genes from a milliliter of blood. The device consists of three chambers: the first separates the cells from the rest of the blood; the second breaks them and obtains copies of their DNA; in the third, a set of DNA markers detects the presence or absence of certain genes, and generates electrical signals to indicate it. The laboratory obtains results in just over two hours. *GATTACA* is just around the corner.

Another bleak possibility predicts that insurance companies could require biogenetic analyzes from those who wish to contract life insurance, denying it (or increasing the fees) to those who have a genetic defect. Examples of this have actually happened, when a man who had suffered from hereditary hemochromatosis and had been treated successfully (the disease was eliminated) was later denied insurance. Sometimes, ignorance complicates things: another company refused to grant life insurance to a person who

---

[170] One nanometer is equal to 10^{-9} meters, i.e. one thousand of one micrometer. 20 nanometers is the size of the common cold virus.

suffered from congenital Charcot-Marie-Tooth disease, which can cause neuromuscular problems, but not death.

The use of biogenetic information for medical purposes may also have undesirable ethical consequences. An example of this was the sickle cell prevention campaign that took place in Greece in the 1970s. The disease is a consequence of a recessive genetic defect, which means that, in order to suffer the disease, the defective gene must be inherited from both parents. As a consequence of the study, many carriers who did not suffer the disease were detected, for they had inherited the defect from just one of their parents. As the list of carriers was made public, these people were subjected to discrimination, because no one wanted to marry them, and they had to find a partner among individuals belonging to the same group, which increased the chances of their children inheriting the disease. This preventive medical campaign should be considered a complete failure, for its result was increasing the number of cases of the disease that should have been prevented.

The fact that people know that they have a damaged gene may have unfavorable psychological effects, such as continuous monitoring, to see if any of the symptoms of the disease occur, with the risk that preventive tests end up causing neurosis.

For all these reasons, in 1989, when he was director of the National Center for Human Genome Research, James Dewey Watson allocated three percent of the budget (later increased to five) to study ethical, legal and social problems that can arise as a result of the knowledge of the human genome. These problems appear all the time and may get even worse in the future.

* * *

We will end with some ethical problems related to biotechnology, frequently discussed in the media and in the most important political and scientific forums.

- *Human cloning*. A cell would be taken from an adult human being and the nucleus is extracted. A human female ovum would be taken, the nucleus also removed, and replaced by that of the adult cell, so that the ovule has the same complete genetic endowment (diploid) as the adult from which the nucleus was obtained. It is, therefore, equivalent to a zygote (fertilized egg), but with the genes of another person. The modified ovule would then be subjected to various chemical and mechanical processes that induce it to divide. From there, the procedure is identical to test-tube babies: the ovule is implanted in a woman's uterus and allowed to develop as an embryo, until a new human being is born, who'll have the same genetic endowment as the adult who donated the initial cell. A clone of that adult would be obtained. The process has worked successfully in various species of animals.

This is another case where the media often misinterpret the issue. They say sometimes, for instance, that cloning would prolong the life of the individual indefinitely. When death approaches, they say, people could make their own clones from one of their cells, and prolong their life in the clone.

This interpretation is delusional. There is no doubt that a clone, obtained from a cell of a human being, would be a different individual, despite sharing the same genetic constitution. The curious thing is that clones have existed since time immemorial and in nature: monozygotic twin brothers arise when a fertilized egg, in the early stages of its development, splits into two parts. Each of the pieces continues to develop independently, giving rise to two different individuals with the same genetic endowment. Each twin brother is a clone of the other. However, despite the surprising affinities between twins, nobody has ever said that both brothers are one and the same individual.

Currently, for ethical reasons, human cloning is forbidden in most countries.

- ***Therapeutic cloning***. This case uses similar techniques to the previous one, but differs in the objective. Everything proceeds the same, until the ovule whose nucleus has been replaced begins to divide. Then, instead of implanting it into a woman, it is allowed to divide inside the test tube until it has about one hundred cells. Finally, these cells are separated and used in experiments. Some of them (the stem cells) are capable of specializing and transforming into cells of any type (neurons, muscle cells, pancreas cells) and could be used to correct diseases of genetic or degenerative origin, such as diabetes or Alzheimer's disease.

The ethical difficulty here affects any experimentation with embryos and has a lot to do with the debate about abortion, which we'll discuss in more detail in the next chapter. The key issue can be reduced to one question: is an embryo a human being? If the answer is yes, it is clear that experimenting with embryos should be forbidden, as any other experimentation with human beings, except in well-determined cases, which should always affect well-informed volunteers.

As for the possible help that embryonic stem cell experiments could provide to sick people, nothing is really known: we just have assumptions. Stem cells can also be obtained from adults, including the sick person, which would eliminate any danger of rejection and would not present ethical problems. In fact, experiments with adult stem cells have so far got better practical results than those using embryonic cells. However, the media, which has taken therapeutic cloning as one of the flags of a misunderstood *progressivism*[171], often ignore scientific news regarding adult stem cells, while airing speculations about embryonic cells.

The use of these techniques for the generation of organs (heart, liver, pancreas...) that could be used to perform transplants is an issue that can still be considered science fiction, but perhaps not for a long time. If the patient's own cells are used to produce these organs, all rejection problems currently presented by organ transplants would be avoided. But if the organs are obtained by therapeutic cloning, rather than from adult stem cells, this hypothetical advance would present the same ethical problems just mentioned.

There is a question, whose answer could help solve these problems: *when does the individuality of a human being begin?* It is clear that the fertilized egg has, from the

---

[171] See next chapter.

moment of fertilization, the same chromosomes (the same inheritance) as every cell of the same being as an adult. There is no solution of continuity in development, except in the case mentioned of identical twins, when the group of cells that comes from a single zygote splits into two or more parts, each of which develops into a different individual. Sometimes, the separation is not complete, so two more or less fused individuals (Siamese brothers) can be born. There is also the case of chimeras, when two different zygotes fuse and develop into a single individual, whose cells belong to two different genetic identities. But these phenomena only occur during the early stages of cell division. In normal pregnancies, they end long before the mother is aware of her condition.

- ***Children on demand***. Using the techniques of genetic engineering and *in vitro* fertilization to select children with predetermined genomes that can be used to provide organs for transplant, or to solve other problems for sick older siblings. In this case there are two ethical problems. The first is general: *can human beings be used as organ factories, as instruments subordinate to another, without their consent?* The second is more concrete: to obtain the child on demand, several eggs must be fertilized. Among the resulting zygotes, the one with the desired genetic constitution is selected. The others are discarded, thus giving rise to the same problems posed by abortion, therapeutic cloning and *in vitro* fertilization.

Man is not God, but he likes to think that he is. This is very dangerous, although in his role as a sorcerer's apprentice he usually ends up ashamed. We all know many examples. Genetic manipulation, the most powerful instrument that scientific research has given us, has a potential for terrible, unprecedented abuse. Therefore, it must be carefully controlled, and its use outside well established limits must be strictly prohibited.

14. How will the fifth level be?

Throughout this book, I have described the evolution of life since the beginning of the universe, through the changes in level that took place on Earth over several billion years, to the present day. We are clearly in the fourth level of life (multicellular beings) and we can see glimpses of the fifth in social insects (hives, anthills and termite mounds) and in human society. Then we have signaled that man is not just one more animal species, but has meant a state change, the passage through a critical point, where biological evolution yielded to cultural evolution its role as the motor of the changes.

In chapters 8 and 9 we noticed that the idea of a fifth level of life is not new, but has influenced literature and philosophy for over two millennia. Next we saw that the advances in information and communication technology have shortened distances on Earth, making possible the construction of a nervous network that could turn humanity into a single body, directed by a nervous system, still decentralized: a headless body. On the other hand, we have seen that man, through the development of genetic engineering, can cause artificial inherited biological alterations. In other words, cultural evolution has not just replaced biological evolution, but is now almost able to drive it where it will.

The time has come to recapitulate all this information, to consider the real possibilities of a new leap in the evolution of life, and to estimate what could be the properties of this fifth level towards which we seem to be heading.

The most obvious extrapolation is considering the fifth level as a super-organism, where individual human beings would play the same role as the cells of our body. For such an organism to be viable, there must be a high degree of cohesion and mutual solidarity between human cells, much greater than what we see today in modern societies. In chapter 11 we saw that there is an evolutionary contradiction between selfish tendencies, favored by natural selection at the lowest level, and altruistic, without which the new level cannot arise. Relying solely on biological evolution, subject to chance, it seems highly unlikely that the fifth level will emerge without the individual human beings being forced to renounce individual reproduction, but this may lead us to an undesirable situation, similar to that described in the novel *Brave new world*.

We have two alternatives: direct manipulation of our biological evolution with the techniques described in chapter 12 (subject, perhaps, to restrictions such as those outlined in chapter 13), and using different control mechanisms related to cultural evolution, which has become predominant in our species. We have seen that both options are far from ensuring the success of the enterprise, that the process may derail one way or another. This is due to the fact that each of the human individuals is free to oppose the objective of reaching the fifth level. In fact, all human beings do it at some time, as long as we put our selfish interests above the common good of humanity.

We now have at least one criterion that allows us to judge human acts, distinguishing between those that will drive us to advance on the path of evolution (towards the fifth level) and those which try to prevent it and would keep us in the fourth. We can, finally, unambiguously define the word *progress*, which throughout recent history has supported all kinds of behaviors, usually contradictory, more or less improperly appropriated by many political parties and many artistic or scientific schools.

Indeed, the word *progress*, by itself, has no meaning. It is necessary to specify where such progress is leading. This term just means *a movement towards some objective that one wants to achieve*, so it makes no sense to use it without specifying the objective. Thus, if we are halfway between A and B and wish to go to B, we progress if we move in the AB direction, but if we want to go to A, progress will make us move in the opposite direction. In this book we are defining progress as any advance in the path that leads us from the fourth to the fifth level of life.

An altruistic human act, which puts the good of others or that of the whole society above its own good, should be considered as an act worthy of the fifth level of life, and therefore marks a real and measurable progress towards that level. On the other hand, any selfish action is a triumph of the fourth level and should be considered as a retrograde act, which opposes the true meaning of evolution.

In the light of this analysis, we can glimpse a possible solution to another big question, an object of permanent debate throughout history: *is there an absolute ethic, which everyone must accept?* Or, on the contrary, ethics is something relative, which depends on each society or each individual? With what we have just said, it is clear that there are at least two different ethics. One of them, the ethics of the fifth level, adopts altruism as the basic moral criterion, the search for the greater good for all human beings. It will, therefore, be absolute, the same for everyone. The other, the ethics of the fourth level, on the contrary, seeks our own good and is based on selfishness. Naturally, it will tend to be relativistic, since the good of an individual does not have to be equal to the good of another.

A particular case of fourth level ethics is the assertion that *what society wants is morally acceptable*. Although, by mentioning society, it looks like the objective is the good of everybody, that may not be true. What society wants is, in reality, what most of its members want, which can be (and, in fact, often is) the sum of multiple selfishness. This is so, especially when those negatively affected by decisions based on *what society wants*, cannot express their point of view to defend themselves. This occurs, for example, in some of the ethical issues related to research in biotechnology, which we have discussed in the previous chapter, or in the issue of abortion.

Using this criterion, let's examine some of the moral issues concerning man just now, or that have concerned us in the near past, and analyze whether there have been advances or setbacks in recent times, and if the first dominate the others. Considering any doubtful human act, we must ask ourselves the key question in criminal investigations: *cui bono?*[172], to decide whether it is a fifth level altruistic act or a fourth level selfish act.

---

[172] Whom does it profit?

1. **Human rights**. They represent a clear advance upon the previous situation. They recognize that every human being, by the fact of being a human being, is subject to certain inalienable rights. Often, when we talk about human rights, when we list them, we imagine that we are speaking about our own rights, what we can demand from others in their relation with us. In itself, this view is correct, but betrays selfish tendencies. It would be much better if we focused the matter from an altruistic point of view, as a function of the good of others. For each of us, the list of human rights should represent the set of things that other persons, by the mere fact of existing, have a right to demand. In other words: we should consider the list of human rights as the list of our duties towards others.

   The first place among human rights is logically occupied by the right to life. From what I have just said, we should consider our duty to respect the lives of others, both negatively (*you won't kill*), or positively (*you will help others to preserve their life*). Both versions are clearly altruistic. In our civilization, the negative form has historically allowed a single individual exception (*legitimate defense*) and two collective exceptions (the *death penalty*, ruled by a legitimately constituted court; and *just war*). We'll talk about war later. Along the twentieth century, the death penalty (a classic form of legitimate defense of society against the attacking individual) has been considered excessive, for society should be strong enough to defend itself against an individual without resorting to the death penalty. The abolition of the death penalty can therefore be seen as an altruistic measure, a breakthrough in the path of progress towards the fifth level. Unfortunately, this advance is not universal, since it survives in the penal codes of many countries of the world.

In its positive form, the duty to respect life moves us to help those in danger of death. This applies especially to that part of the world's population threatened by hunger and other urgent needs. It is often said that rich countries should devote 0.7% of their GDP to help poor countries (a commitment that almost none fulfills), and it is frequently added that just a small proportion of what world spends on (say) weapons, oil, etc., would solve the problem. We pass our time protesting against governments; wouldn't it be more effective, when these governments won't do anything, if private persons did it? How many problems could be solved, if each of us gave a reasonable percentage of our net income to serious and trusted non-governmental organizations (Catholic missionaries, for instance), to help those who are hungry? A very large percentage wouldn't be necessary. Probably, if everyone contributed at least 1 percent, there would be more than enough. Look at yourself and your finances, and think if you could live with 99 percent of our current income, or if that small loss of personal comfort outweighs the loss of many lives.

2. **Equality**. Since the French Revolution to the present day, the idea that all human beings are equal has progressed unstoppably. In itself, this idea should be considered an advance toward the fifth level, as the belief of being better than others because of class, sex, race, religion or education

is obviously a selfish attitude. We must be careful, however, to make clear what kind of equality we are speaking about: rather than functional equality, we should be dealing with equality in intrinsic value; equal opportunities, not than equal numbers. If we observe the change from the third to the fourth level, as in Chapter 7, we'll see that *union differentiates*[173]. The most different cells are those belonging to the same multicellular organism. A nerve cell (a neuron) and a muscle cell differ from each other much more than an amoeba and a single-celled alga.

Therefore, we must conclude that the position of radical feminists, who insist on total equality of function between men and women, may be justified as a reaction to the previous situation, when there was a difference in treatment and in rights between the two sexes, but it can be taken to the point of surpassing reasonable limits, in the light of the parallels just mentioned. Will someone demand that 50 percent stevedores be women? Or that a different cut-off note should be applied to boys and girls for university access, so that the number of students of both sexes in all studies be identical?[174]

No one will ever believe that equality between human beings means that everyone must have the same height, the same skin color or the same profession. However, modern

---

[173] This is how Teilhard de Chardin expresses it in *The phenomenon of man*, book four, chapter II.1.B, *The personalizing universe*.

[174] A Cuban university applies this absurd system among those who wish to study computer engineering.

teaching methods insist that all children should receive the same education and achieve the same results. If functional equality is taken to the extreme, we shall find that children without innate physical abilities must make superhuman efforts to pass a Physical Education course (sometimes with tragic results), or that children unable to detect differences in musical tone cannot proceed with their studies because they cannot pass Musical Education. Current teaching systems have many deficiencies. To quote one, they should be more differentiated than they are now, adapt to the functional individuality of each student. Perhaps technology will make this possible in a not too long future.

We live in a more egalitarian society than former ones, which undoubtedly constitutes progress. However, we must be vigilant, so that an excess of egalitarianism does not push us back. Paraphrasing Aristotle, *virtue is in the middle point*.

3. **Private property**. Man has an innate instinct of private property. We need to have some possessions that ensure our life and the exercise of our profession. However, this trend shouldn't exceed certain limits, such as yielding to the wish to seize more and more things, far surpassing our needs. Greed (one of the classic capital sins) has been recognized since antiquity as a selfish attitude, contrary to the good of society. Misers have been ridiculed in many classic and modern works[175].

---

[175] Plautus, *Aulularia*; Molière, *L'Avare*, etc.

Given this situation, many authors of utopias believed that the problem can be solved by forbidding private property, which they consider as the origin of all evils. We have seen this in Chapter 8, in Plato's *Republic*, Thomas More's *Utopia*, or Marxism. But this attitude confronts an innate tendency in man. In fact, Marxism does not require the elimination of all types of property, just production goods.

In our current state of development, a balance should be sought between acquisitive tendencies (which are selfish when certain limits are surpassed) and altruistic tendencies. This limit could be achieved if individuals who accumulate a lot of wealth should keep a reasonable proportion for themselves and make the rest available for those who have less than necessary. State governments have played a role in the redistribution of goods through taxes, although they are not exempt from criticism, if they squander or misuse public funds, which incites taxpayers to avoid participating in common welfare. We have too many examples of this type of behavior, and are no longer surprised by it.

4. **Slavery**. For most of mankind's history, it was considered correct if a human being had a right of life or death over another, and would take advantage of his work, giving nothing in return, apart from survival maintenance, and even this could be eliminated if desired. Our civilization has not escaped this scourge, obviously selfish, until a recent past. In the sixteenth century, on the occasion of the discovery and colonization of America, a trend emerged in Spain to defend the rights of the natives and oppose their enslavement, not always successfully. However, the

enslavement of the black peoples of Africa was generally admitted throughout the West. As a justification, their being human beings was questioned, and they were considered as animals.

In the mid-nineteenth century, this situation became unsustainable, so that slavery disappeared in a few decades, at least openly, in all the countries of the West. In 1845, with the promulgation of the *Aberdeen Act*, the United Kingdom declared slavery outlawed, authorizing its ships to attack, even in jurisdictional waters of other countries, any vessel, whatever its nationality, engaged in, or suspected of engaging in slave trade. In the United States, the eradication of slavery became one of the reasons for the Civil War. Curiously, the argument that slavery should not be forbidden, but left to personal conscience, was used during the prolegomena of that war. If some persons are not conscience stricken at having slaves, it's their problem, and others shouldn't butt in. This same argument has been used to defend selfish behavior of different types, as we'll see.

It's clear that the abolition of slavery benefits humanity in general and only harms those who wish to take advantage of others for their own benefit. It is, therefore, a progress, a clear advance of altruism against selfishness. This is so, although it has been said that the reasons that led the United Kingdom to take active measures against slavery were not so altruistic: the possession of India assured them an almost inexhaustible workforce, so they didn't need black slaves and didn't want other countries to have them. Can an altruistic argument be used to hide selfish ends? Perhaps, but we've already signaled that in every human act both tendencies are inextricably mixed. It must be recognized, in any case, that the abolition of slavery meant an unquestionable long-term victory of altruism over selfishness.

5. **Racism and exacerbated nationalism**. Both cases are closely related: they deal with asserting the superiority of a race, a nation, a culture, over others. It is, therefore, a collective selfish attitude. It's curious that, nowadays, racism has an evident bad press, while nationalism is more tolerated. The defense of the identity of an individual or a group is reasonable and convenient, because differences must be cultivated, but their violent imposition exceeds the limits.

    Although, in theory, nobody likes to be called racist, this type of attitude arises where least expected, so it's necessary to keep the maximum vigilance to avoid it. In any case, it should be borne in mind that some behaviors that the press describe as racist, sometimes are not, either because the term is used as a pretext or as an insult, or because it mistakes racism with cultural shock that, although related, is different.

6. **War**. For most of history, war has been a well-regarded social activity. Only recently has a pacifist movement emerged, and spread throughout Western Europe during the twentieth century. Undoubtedly, war is in principle a selfish attitude, being a violent way of solving problems between

societies, so its loss of prestige is an unquestionable advance towards the fifth level.

However, as in other cases, care must be taken of not taking too far the pacifist stance. Just wars do exist. No one has dared to assert that the Allied powers shouldn't have declared war on Hitler, when he proceeded relentlessly with his plans to conquer the world for racist reasons. No one dared to criticize the United States when they launched an attack against the government of Afghanistan, because they refused to hand them Ossama Bin Laden, instigator of the attacks on the twin towers. The same did not happen, however, with the second war in Iraq. Finally, the pacifists themselves called out for NATO intervention in the wars of the former Yugoslavia, first in Bosnia-Herzegovina, then in Kosovo, in the face of the danger of a genocide. Legitimate defense, usually admitted as an exemption, can also be applied at the social level.

7. **Abortion**. This moral problem has grown enormously since the second half of the twentieth century and continues today. From a scientific point of view, the question of the origin of human life was solved long ago. The biological consensus can be summarized in the following three statements (check any biology text, or any encyclopedia):

- Every living being generated by means of sexual reproduction begins its life with the fertilization of the female ovum by the male gamete: with the formation of the zygote. At that time a new being of the same species as its parents is generated, whose genetic endowment (DNA) is different from that of its parents and from any other living being of the same species (except for identical twins). This new living being will retain the same genetic endowment from that moment until its death. This is the reason why fertilized eggs of sea turtles and other endangered species are protected, because they enclose individuals of those species.

- In all species of living beings that do not suffer metamorphosis (this includes all reptiles, birds and mammals and, of course, man) there is no solution of continuity in development, from the zygote to death. The phases usually distinguished in the development of human beings (embryo, fetus, newborn, child, adolescent, adult and elderly) are arbitrary and without discontinuities. Even childbirth is not a discontinuity, for anatomically it just consists in cutting a blood vessel (physiologically it has other effects). In all these phases, from beginning to end, there is no doubt that we are dealing with the same individual.

- In all placental mammals (including man), the first phase of the life of a new individual takes place within the mother's body. The pregnancy is equivalent and replaces the development in the egg, which takes place outside the mother in reptiles and birds. In all cases, maternity takes place at the time of fertilization, not at birth, which corresponds to the breaking of the eggshell.

This is the scientific consensus of biology, accepted by all biologists. So why are there abortionist biologists? Because they follow the relativistic ethics, asserting that *everything society wants is morally acceptable*. Because they think that the decision in this respect has nothing to do with science, just with laws.

Before the decision to perform an abortion, a woman should try to have all the information, which sometimes appears to be hidden, so that she doesn't change her mind. There are many educated people who do not know elementary biological concepts, as the fact that *an embryo is a living being that belongs to the human species, and is different from the body of the mother*. Then, along the lines of what we have said above, she should ask herself whether abortion is a selfish or an altruistic act.

It is ironic that abortion supporters call the defenders of life *reactionaries* and describe themselves as *progressive*, when what they have done is bring society back, to the state it was two thousand years ago, in the Roman Empire, where abortion was legal without restrictions and infanticide was allowed up to 24 hours after birth. A relic of this is our current legislation, which does not consider a newborn child a legal person until 24 hours after birth. It is also curious that some of the arguments of the advocates of free abortion are similar to those used in the past by the advocates of slavery.

8. **Suicide and euthanasia**. Although in Rome, Japan and other cultures suicide was considered a worthy exit from life, in our civilization it has usually been considered as an act of cowardice, a way of avoiding one's responsibilities,

the extreme triumph of selfishness over altruism. Suicides leave the fight to get rid of pain or dishonor, without taking into account that, by choosing to die, they are denying support to those who in the future might have needed their help.

Recall the case of Kevin Carter, a famous photographer who received the Pulitzer Prize for the photograph of a famished girl who, observed by a vulture, was trying to reach the food center of an African village. Carter was criticized for having waited twenty minutes to take the picture, instead of helping the girl. Some believe that those criticisms pushed him to suicide. He was 33 years old when he died. Assuming that this was the real reason, wouldn't it have been better if he had spent the rest of his life helping other people? Wouldn't he have compensated the shame that, perhaps, he couldn't overcome? Can it be doubted that this attitude would have been more altruistic and less selfish?

A similar case is active euthanasia, which is usually presented as if it were an act of mercy towards the sufferings of a sick person, as an aid to committing suicide for someone who wants to die; but one often hidden reason may be to release their family members of their responsibility, by applying the utilitarian conception of the person that has spread throughout the world in flagrant contradiction with the doctrine of human rights. In the case of childhood euthanasia, it is even worse, for this justification cannot even be adduced, as children are killed without their consent.

We mustn't mistake active euthanasia, which is the conscious act of killing a person, with passive euthanasia, which consists in dispensing with *therapeutic cruelty*, defined as the lengthening at all costs, by artificial procedures, of the life of a person when the time of death is coming. Again we can see the intrinsic contradiction of our society, hesitating between the fourth and the fifth level. While defending the active death of those who could live but don't want to, sometimes it prevents the death of those who have no longer a future. Often this cruelty just represents the fear of the doctor to be subject to a lawsuit, to be accused of a defective treatment; in other words, a selfish attitude, although not as bad as that of the lawyer who systematically advises relatives to sue doctors so as to get economic benefits from the death of their relatives.

9. **Uncontrolled consumption**. Modern economy seems convinced of the need for a permanent positive growth to keep the standard of living of humanity. The defenders of *zero growth* seem to have lost the game since the abrupt collapse of communism, which took place in 1989, and the subsequent triumph of capitalism. However, sustainable positive growth is a fallacy. Any mathematician knows that a constant growth, however small (say, 1 percent) quickly leads to an exponential increase. But natural processes are never exponential: although some systems grow at the beginning in an almost exponential curve, they actually follow a logistic curve that tends asymptotically to a stable maximum value. See figure 14.1.

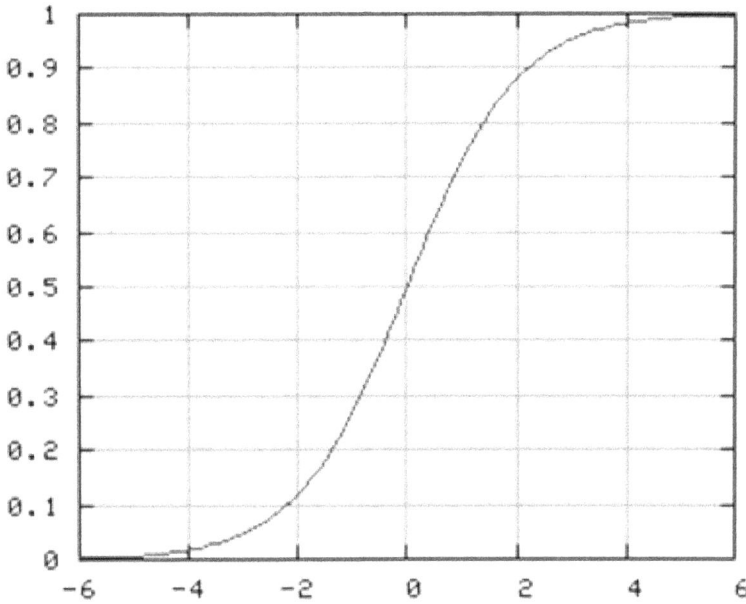

Figure 14.1: Logistic curve

An exponential growth such as modern economists advocate will inexorably lead to the depletion of natural resources. We have here a purely selfish attitude of the current generation of human beings, which seeks its own benefit at the expense of our descendants.

Neither modern capitalism, with its obsession with exponential growth, nor communism, which turned Eastern Europe into the most polluted region of the globe, with extremely harmful effects in the Chernobyl accident, are in the proper path. We have seen in Chapter 11 that economic liberalism without palliatives tends to favor selfishness over altruism. We must find a third way that will take into account the interests of all humanity, present and future. We must modify our economic theories to make zero growth possible in the use of natural resources. Only in this way we'll be able to guarantee the indefinite conservation of economic structures until some important technological advance takes place, such as nuclear fusion.

* * *

Let's recap: from the previous analysis it is clear that, in some of the moral issues of our time, the altruistic position has made progress, while in others there have been setbacks and a selfish attitude dominates. As usually, we are a mosaic of the fourth and the fifth level, of good and evil. But I have the feeling that, as a whole and on average, we live in a time of setback. Rather than approaching the fifth level, we are moving away from it. To verify this, we can study the statistics that show an increase in the number of crimes, in a proportion much higher than the increase of the population. We can contemplate the global loss of values, of the ethical meaning of life, the generalized predominance of short-term enjoyment reflected in uncontrolled consumption, the progressive loss of value of the human life, which tends to acquire a purely utilitarian meaning. In all these features selfishness (individual or social) unequivocally dominates over altruism.

If we now look at social organizations, political parties and the like, we see in each of them a similar mixture. Some place more emphasis on some values, others on different values, but there is hardly any that can be said to adopt always, in all the issues we have discussed, the defense of the altruistic attitude against the selfish.

I fear that this evolution will lead us directly to a catastrophe that has been announced by most of the great philosophers of history in the twentieth-century: Spengler (the title of his work, *The Decline of the West*, says it clearly); Toynbee, for whom our civilization has passed the collapse phase, which would have taken place during the two world wars; and Sorokin, who detects in history a clear alternation between idealistic and materialistic epochs, with catastrophic collapses at the end of the second, and believes that we are quite advanced in the race towards social self-destruction.

* * *

We have defined *progress* as a measure of humanity's advance towards the fifth level, but this does not mean we believe in *the myth of indefinite progress*[176]. Remember the Toynbee simile, which compares the course of history with a vehicle whose wheels, while just rotating, make it move forward. Cultures and civilizations rise and collapse; but humanity, as a whole and in the long term, advances. From the fact that I think that we are moving towards the fifth level, does not follow, in the first place, that we are going to succeed (we can spoil the attempt), nor that it will be a simple path, without repeated ups and downs, and setbacks. I said a moment ago that we are in a time of setback. Tomorrow things can change and we can start a time of progress. In this I differ from Teilhard de Chardin, who foresaw an accelerated and convergent advance towards the Omega point. I don't see signs of it. In my opinion, good and evil (altruism and selfishness) progress approximately, on average, with the same speed. Teilhard himself admits that the struggle between good and evil can increase in the last times: *But there is another possibility: Obeying a law of which there are no exceptions in the past, evil can grow along with good and reach paroxysm in the end in a new way*[177].

---

[176] See chapter 8.
[177] *The phenomenon of man*, book four, chapter III.3.

Let's look at an example, which I myself witnessed through television. In the nineties, a terrible civil war took place in Bosnia-Herzegovina. One night, the news showed pictures taken in Sarajevo. People were seen walking quietly down the street. Suddenly, a sniper began firing from the roof of a nearby house. Immediately, everyone started running to escape from the bullets. A lady in her seventies was hit by gunfire and fell to the ground groaning. The scene was enough to lose faith in the human species.

Suddenly, hearing the groans of the lady, two or three young people who were fleeing stopped, came back, picked her up and took her away, putting their life in danger to save a wounded neighbor. The scene was enough to restore faith in the human species. We are capable of great evils, but also of great heroisms. Evil is growing, for we have more means to do it, but the same applies to good. The struggle between good and evil, altruism and selfishness, the fourth and the fifth level, will always be with us. The final outcome of the contest is not clear, but there are still hopes.

I have said that the process can derail. Does this mean that I think the fifth level of evolution can be unattainable? Not necessarily. But I fear that we must assert that its subsistence is practically impossible in a world subject to the passage of time, because time allows us to act and move against the path of evolution, if we wish. Time and the stability of the fifth level of life are incompatible. The latter can only be viable if, at the very moment in which it begins to exist, it leaves time forever to pass into eternity or a different mode of time.

There is another reason that leads us to demand this transcendental step: even if we could conceive of a perfect society without the indicated defects, although the people in that society were able to change their essence and become really capable to belong to it, even so, their stability could not last forever: time itself acts against it, for the universe is moving towards its end, like a runaway horse into the abyss. In the long run, the thermal death of the universe will come, either at absolute zero, if the cosmos is open and its expansion continues indefinitely, or at the fiery temperatures of a new cosmic egg, if it were closed. Nothing that man can do can stop this process, because the second principle of thermodynamics and the law of gravitation are inextricably linked in the warp of the universe where we belong. It's impossible to change this from within. Consequently, our imaginary perfect societies, More's *Utopia*, the Marxist society, the scientist or the ecologist utopias, if they come into existence, would be doomed to die, like every one of its members.

From a strictly rational point of view, the end of the universe makes every effort to reach a perfect society futile, unless we allow for a way out, a pass to immortality. Otherwise, all our efforts will be useless, because no one will remember them in a universe that will have disappeared forever. From the hypothesis of total destruction in the long run, only quietism (refraining from any effort) or selfishness (refraining from any sacrifice, except to obtain a greater good for oneself) are rationally justified.

The sacrifices we make are always aimed at ensuring some good for ourselves or for others, or to improve the memories we leave after our death. But if everything, without exception, will disappear without a trace, without leaving a memory in some consciousness;

if we are eventually condemned to oblivion, why bother about the suffering of others, why alleviate it, if the entire universe will forget about it, it will be the same as though it had never happened?

This way of seeing the world is spreading on the wings of atheism in our decadent civilization. Many critics of art, film, literature, only praise nihilistic, desperate works, while showing their contempt for anyone who tries to present an optimistic view of the world where we live.

I completely disagree with this vision of things. For me, life is wonderful; I don't want to run away from it; I don't find it banal, distressing or boring. On the other hand, the nihilistic stance can be, in large part, a pose. In my opinion, the facts show that, in practice, there are very few philosophical materialists[178] in the world. Consciously or unconsciously, we all believe that something can be saved from absolute destruction: we all hope for immortality.

* * *

After this long journey, we can try to describe the properties of a being that belongs to the fifth level of life. We'll start asking ourselves if there can be several.

Modern technology has evolved to such an extent that the Earth has shrunk. Today it's as easy to get from one end of the planet to

---

[178] Don't confuse *philosophical materialism*, which asserts that only matter exists (in a wide sense of the word *matter*), with *methodological materialism*, which is a part of the scientific method and assumes, when an experiment is performed, that only natural causes will act, excluding supernatural interventions, but not denying their existence.

the other as it was for our near ancestors to travel to the capital of their province. Whether we like it or not, the society of the future is bound to be global, to cover the entire Earth (this is what we call *globalization*). In the future, we will undoubtedly become a unique society. There will therefore be just one (if any) being of the fifth level.

The tendency towards social unification is so great that we can go further and make a prediction: if there were extraterrestrial intelligences, and if we'll get in touch with them, they will all come together (and with us, if we survive), so that, after a long evolution, perhaps billions of years, a unique cosmic society will be formed. If this happens, that's where the fifth level would come to exist.

What will be the role of a concrete human being after becoming a cell of this being? Is there any kind of society that, right now, can give us an idea of the fifth level of life?

The mass, the crowd, and other disorganized pluralities can be eliminated as an element of comparison. We must reduce our study to well-defined groups, provided with clear internal organization, if we want to follow the clues provided by the examples drawn from other levels of the evolutionary movement.

We can also forget those societies that, although well organized, achieve their unity through coercion. We know that a referendum, held in a country under a dictatorial government, almost always approves near unanimously, but we should not confuse unity with unanimity, especially if the latter is coerced. Must our perfect society be a dictatorship? This would lead us to conclude that, after all, we are not going towards utopia, but towards dystopia. On the other hand, a society based on tyranny cannot last forever. Sooner or later, human freedom will rebel against oppression, and tear it down.

Throughout history, the dilemma of the relative importance of the individual and society has been raised once and again. For philosophical materialism, society must be more important: it has the advantage of size and duration. The problem appears to be equivalent to the comparison between the relative importance of the animal and each of its cells. In this case, there is no doubt: the cell must sometimes sacrifice its own life for the sake of the higher level being. If it refuses to do so, if it acts on its own, it triggers a cancer that, in the long run, will destroy the community, together with the individual that originated the rebellion and its descendants.

If, on the contrary, we believe that all individual human beings are immortal, the problem changes: the comparison with the cell and the animal is no longer applicable. In that case it seems clear that the individual should be more important, since he is immortal, while none of our societies is. The advantage of duration is on our side.

But now we face a situation where both terms of the comparison, human cell and being of the fifth level, can be simultaneously immortal. If our utopia is true, if the fifth level of life manages to cross the barrier of time, we, its cells, will go with it outside our time. Individual immortality would be ensured at the same time as the collective. Then who is more important?

The dilemma is false. Cells are important because the whole is, and vice versa. Neither individualism nor collectivism give us the solution to the problem. We must find a third way.

There is a type of society in today's world that approaches, albeit imperfectly, to the combination, the incorporation of the good of the individuals and the community. This is the family, where each and every member occupies a different and irreplaceable place; a place, so to speak, organic. No one is interchangeable. No one can be confused with the others. No one is unnecessary: a family lacking one of its members is incomplete, like a mutilated body lacking an organ. Strong dependency relationships are established among all members.

The unity of the family, the joint solidarity of its members, can be, and sometimes is, very great. We can find in the family the greatest examples of altruism. It is also there where selfishness stands out most unpleasantly. The conclusion of these considerations is evident: the family is the closest social structure, here and now, to the fifth level of life.

The family, however, is not a perfect society. Human beings cannot renounce forever, in this world, to selfishness. Also the family lives in time, which makes it unstable. The children grow up, leave the home and form their own families. However, we know perfectly well which is the link that holds this society together; which is the force that stands most clearly when family members behave as we might expect the cells of the fifth level being to act. *That force is love.*

In our days, when we talk about love, we usually think about sexuality, or perhaps another form of love, the affection we feel for our closest people. In both cases, love is usually identified with a more or less tender feeling experienced towards another person or, by extension, towards animals or objects. But that is not the kind of love that holds the family together.

Our western culture uses to identify love with a passion[179], with something we receive from outside. This is due to the importance attributed to *falling in love*, which takes place at the beginning of the mutual relationship of the human couple. When we say that we are in love with someone, we are describing feelings: a mixture of sexual attraction, aesthetic pleasure and admiration for the qualities of another person. But true love, which is the basis of the stability of a family, is quite different. Love is not a feeling. Rather than a passion, it is an action, something that comes out of ourselves, consciously, voluntarily and freely. It is an act of will, an irrevocable decision.

This tragic error of our civilization and our time, which is the cause of so many family failures, is not shared by other men, other cultures, other periods of history. The American writer Lois T. Henderson tells the case of an Indian boy who moved to the United States to study. After a few years, he announced that he was going back to India to marry a young woman that his family had chosen as his wife, whom he barely knew, and would bring her to live in America. Lois Henderson was amazed. What if they weren't compatible? How could he intend to take her so far from her family and her friends?

"Don't worry," said the boy. "I will love her enough to make up for everything."

---

[179] From the Greek *pathos*, what is felt, experienced or suffered.

"But how can you be sure?" asked the writer.

"Because I have decided," he replied.

He didn't say "I'll try to love her". He said "I will love her." He did an act of will... Now, twelve years later... I see happiness on their faces. If a marriage has been happy, if a wife has succeeded, they have[180].

In recent decades, the concept of love has degraded even more, becoming confused with sexuality: the term *making love* has become a synonymous with the sexual act, although originally it did not have that meaning. At the same time, in one of those contradictions so frequent in the human species, sex has been disconnected from love and fidelity and the family, since the freedom of its practice has been risen as a flag of the sexual revolution. This has had such dire consequences as the proliferation of AIDS and the difficulties encountered to stop it, because if someone proposes safe measures, based on fidelity rather than promiscuity, one must face the automatic anathema of the media and of *progressive* politicians.

Note that, when I'm talking about the family, I don't mean any specific type. Like any human or natural construction, the family also evolves, but it retains a permanent hard core that resists changes: it is the basis of the conservation of the species and of society, it involves the origin of the life of every human being, and it provides the environment in which it can best function and grow. Children who have been raised in failing families or in public institutions are often marked for life; they are aware that they are missing something.

---

[180] *Daily Guideposts*, 1981.

Throughout the last century, we have witnessed a frontal attack on the family, usually associated with political tendencies that call themselves *progressive*. However, if the analysis we have done here is correct, if the family is the current form of society more similar to the fifth level of life, the next stage of evolution, we must conclude that the defense of the family is a part of real progress, while the attacks against it, wherever they come from, are a backwards thrush, which seeks the triumph of selfishness over altruism and the destruction of the best in human beings.

* * *

In short: the being of the fifth level, if it must be viable and consistent with the progress of evolution, must have the following characteristics:

1. *Cellular structure*. Individual human beings will be the cells that join together to form a higher level entity.

2. *Cell differentiation*. Each member will take its own place and play a unique and irreplaceable role.

3. *Cellular dependence*. Life outside the fifth level will be impossible or not worth living.

4. *Solidarity*. The cells will, freely and voluntarily, transfer their own will to the superior good of the whole. Altruism must finally defeat selfishness.

5. *Unity*. The force that will link cells with each other and ensure the stability of the whole is *love*, understood as an act of will, the reason behind the action of each and every cell, rather than a passion they experiment.

6. *Uniqueness*. There can only be one being of the fifth level.

7. *Immortality.* Time and the fifth level of life are incompatible.

A pending problem remains: will all the cells belonging to the fifth-level being (all the past, present and future human beings) freely renounce selfishness, adopting the superior good of the whole as their sole objective? Or will some of us refuse, maintaining selfishness as our fundamental objective, without accepting altruism?

Unfortunately, the fact that human beings are free to choose implies that we can choose evil. Therefore, we must allow the possibility that the second option can exist: that some of the cells of the fifth-level being, unable to accept a permanently altruistic behavior, may encyst themselves and cut off their relationship with the others. We don't know what their life will be like in those conditions, but it certainly must be a real hell.

We'll end the book by trying to see that all this is not a mental imagination; that this being is really possible; and that, with another name, humanity has known it for a few thousands of years.

15. Does the fifth level exist?

In the late nineteenth century, the Scottish anthropologist Sir James George Frazer (1854-1941) presented in his famous work[181] the theory that primitive man first went through an *era of magic*, when he tried to control physical phenomena through enchantments and spells; when he realized that these procedures did not produce the desired results, he assumed that those phenomena are under direct control of supernatural beings, whom it was possible to ingratiate through sacrifices and prayers, thus passing to the *era of religion*; finally, in the modern world, we are about to enter a third era, the *era of technique*, when the desired control is obtained through the knowledge of the physical laws that govern the universe.

More than a century after *The Golden Bough*, history scholars don't feel inclined to accept Frazer's conclusions. Today there is a greater number of historical, prehistoric and anthropological data that contradict them, as they seem to prove that the religious phenomenon is almost as old as man. It's now known, for example, that the Neanderthal man buried his dead along with various useful or symbolic objects, which shows that, already in that remote antiquity, there was some kind of belief in the continuity of human life beyond death. A life that, undoubtedly, was not imagined very different from the life that those hunter-gatherer peoples knew.

---

[181] *The Golden Bough*, 1890.

Beliefs in the afterlife vary widely between different religions. Some, those of relatively uncivilized peoples, maintain the old conception of the other life as a mere extension of the present: the ancient Slavic religions, those of the Baltic peoples, possibly that of the Celts, can be framed within this group.

In ancient Egypt, the problem of survival after death seems to have become an obsession. At first it was only the king (the pharaoh) who, as a representative of the gods, could achieve immortality. Little by little the privilege was extended to other people, until, during the second millennium BC, the *democratization* of the other life was complete.

The Egyptians considered the preservation of the body an indispensable requirement for immortality[182], hence the complicated embalming operations they performed. Despite their efforts, however, the results not always were satisfactory. Fortunately, the bodies of the deceased could be replaced by a statue or a mask representing them. In the worst case, it was enough if their name was written on the walls of the grave.

The dead must pass a trial before a court of forty-two gods, chaired by Osiris, lord of the underground world. Their good and evil actions were compared in a ceremony where the deceased's heart is weighed on a scale. The other dish holds an eye or a pen. While this operation is being carried out, the dead persons offer their defense, by denying having committed any sin, and begging their heart not to bear witness against them[183].

---

[182] This idea was shared by other peoples, as the old Peruvians, for instance.
[183] *Book of the dead*, conjuration CXXV.

The other life was considered to be a continuation of our present situation. The deceased are afraid of having to work in the afterlife. To avoid this, figurines of slaves and workers are placed in the grave, who must replace them in the performance of the tasks they'll be assigned in the afterlife. In general, their future existence is expected to be happy. At night, the dead leave their underground dwelling, and with a light they walk through the sky: they are the stars. Family members meet again. They can devote themselves to hunting, fishing, sailing, turn into birds and fly, go with the sun on its daily trips.

The Mesopotamian version is quite different. The country of the dead is a land of shadows and is located underground. The afterlife is not enviable: the dead must feed on mud or the remains of dishes thrown as rubbish, without ever seeing light, unable to return to the world of the living, which they miss. To stay in the kingdom of darkness is a terrible fate, even for the gods.

The only way to avoid this fate, reserved for the vast majority of human beings, was to achieve immortality, understood as the indefinite prolongation of life on Earth. This privilege, enjoyed almost exclusively by the gods, was granted by them to just two mortals: Utnapishtim and his wife, the heroes of the Sumerian-Akkadian myth of the flood. The epic of Gilgamesh tells how the king of Uruk of this name, horrified by the death of his friend Enkidu, embarks on a desperate search for immortality. To do this, he seeks the advice of Utnapishtim, whom he finds after many adventures and dangers. But the old man disappoints him: *When the gods made men, they were destined for death; eternal life was reserved for themselves*[184]. However, he is offered an alternative, an open path: *Try not to sleep for six days and seven nights.* This seems to indicate that the Mesopotamian religion admits the possibility of achieving immortality through rites of initiation, typical of mystery religions.

The early Greco-Roman conception of the netherworld is similar to the Mesopotamian. The deceased, after crossing the Styx lagoon in Charon's boat, has no hope of ever returning to the world of the living. Their existence in the underground kingdom of Hades is sad: they are hardly more than a shadow.

The Greek religion admits two exceptions to this rule: one is reserved for heroes, who escape death and are miraculously transported to the islands of happiness (the Elysian Fields). The second possibility is accessible to all, in principle. As in the Mesopotamian case, it depends on the rites of initiation in various mystery cults, which assure the adepts eternal happiness. There were three main cults of this class in the religion of the Greeks: the Eleusis mysteries, related to the cult of Demeter; Orphism (a name that comes from the hero Orpheus); and the Dionysian rites, so called in honor of the god Dionysus or Bacchus.

The idea that the dead are relegated to a netherworld, dark and unpleasant, also appears in other early religions, such as the ancient forms of Japanese Shinto. However, exceptions are often allowed for warriors killed in battle, for whom a special paradise is reserved, where they can enjoy the pleasures of their trade. An example is the Walhalla of the German-Scandinavians and the religion of the Aztecs.

---

[184] Words from Siduri to Gilgamesh, confirmed by Utnapishtim.

In India, the Aryan peoples introduced their typically Indo-European religion (*Vedism*), which incorporates eschatological beliefs similar to those just described. But since the beginning of the first millennium BC, it was claimed that some human beings are able to reincarnate after death. Over time, this idea became the center of Brahmin-Hindu theology and was also accepted by the other two great religions that emerged in the subcontinent during the 6th century BC: Jainism and Buddhism. According to this worldview, most living beings are subjected to a series of successive reincarnations, where every death is followed by a new birth, the nature of which depends on the actions performed in the previous life (*karma*). Salvation consists in freeing yourself from this chain through one of the paths accepted by each of the three religions mentioned above.

During the first half of the first millennium BC, the religion of India shifted towards atheism or impersonalism. According to this conception, the eternal absolute element is *Brahman*, which is not a personal God, but an entity that represents the force that gives efficacy to sacrificial rites, among other things. When the vital principle of man (the *atman*) reaches salvation, it fuses with *Brahman* and is depersonalized. For the Hindu philosophers who composed the Upanishads, the *atman* and the *Brahman* are different manifestations of the same universal principle. Their religion is, therefore, pantheistic.

Late Hinduism, which was fixed during the second half of the first millennium BC, as a result of the fusion of orthodox Brahmanism with theistic popular currents, admits the existence of a personal supreme God (Siva or Vishnu, depending on the sect). Brahminical pantheism now becomes panentheism: God is everything in a sense, but he is also an individual, an independent and transcendent person in regard to creation. The remaining elements of the Absolute Truth (living beings, nature and time) emanate from Him, have the same qualities, but are clearly inferior to God in quantity.

The acceptance of a personal God moved Hinduism to add a new path of salvation to those known for centuries (*knowledge* and *meditation*). This is the path of *devotion*, which seeks fullness in the living being's loving union with Absolute Truth, through the direct and mystical experience of God. The devotee who surrenders to God is free from material contamination and the consequences of his actions. When he dies, he won't reincarnate, but unites with the Supreme Being.

Jainism is an extremely regulated and detailed religion, based, like all those in India, on the doctrine of reincarnation caused by the consequences of human acts. The path of salvation in this religion is very long and difficult. In their conception of the world, there is no place for a supreme God: liberated living beings are all equal to each other. The fullness of being consists in obtaining pure knowledge, which gives eternal happiness.

For Buddhism, another atheistic religion in its original form, the liberated state is *Nirvana*, which is defined negatively as the extinction of all pain, all action, all suffering, all physiological phenomena, all psychic processes. But it is expressly denied that Nirvana be identical to nothing. They don't even admit its identification with unconsciousness. It is, rather, an imperceptible, infinite and resplendent consciousness.

In China, primitive religious forms believed in the existence of two souls in each human being. The first, belonging to the *yin* category (related to the universal feminine principle), begins to exist at the moment of conception of the individual. The second, corresponding to the *yang* category (the universal male principle), is created at the moment of birth. With death, both souls separate. The *yin* soul stays in the grave, next to the corpse, and must be fed and honored by the dead person's descendants. The *yang* soul (originally believed to be reserved for people of high social class) has to travel a difficult and dangerous path to reach the abode of the God of heaven, where it hopes to be admitted as a heavenly guest and live a life similar to our own.

Confucius, the great Chinese philosopher of the 6th century BC, whose ethical rules became, centuries later, an official religion, does not have much to add to the ideas commonly accepted in his time about the other world. On the other hand, Taoist doctrines, which emerged during the second half of the millennium, introduced some novelties. Over time, this religion, Taoism, developed esoteric and alchemist techniques that allowed the adept to achieve long life and higher degrees of immortality. The *spiritualized bodies* of ascetics would possess extraordinary properties, such as the ability to fly and pass through material objects. There are, according to the Taoists, several categories of immortals: those who ascend to heaven forever, those who remain indefinitely in this life, and those who seemingly die in the body, while their spirit goes to the islands of the blessed, where the immortals reside in palaces of gold and silver.

Mazdeism is another great universal religion, which emerged in Persia during the first millennium BC. Its founder was Zoroaster or Zarathustra, whom most modern researchers tend to consider a historical character. Eschatological predictions, individual and universal, constitute a fundamental part of the doctrine of this religion, whose name derives from that of its supreme God, Ahura Mazda, *the Wise Lord*.

Zoroaster's religion was partially dualistic in its origin and exaggerated this character over the centuries. Initially, the Wise Lord was accompanied by two twin spirits, Spenta Mainyu and Angra Mainyu, who were equal in everything, but chose different destinations: Spenta Mainyu chose good and life and remained faithful to Ahura Mazda; Angra Mainyu preferred evil and death, separating from Him forever. Since then, human beings are offered the same choice. They can choose the *party* of the wise Lord (i.e. accept Zoroaster's guidance) or the opposite field. This decision, fundamental for everybody, decides the fate of the soul after death.

After remaining for three days near the corpse, when it begins to show signs of decay, the soul starts a long journey. Shortly later it meets its *daena*, a term on which not all researchers agree, that could be a personification of conscience. Those who acted well, see the *daena* as a beautiful young woman, about fifteen years old. For the wicked, on the contrary, she appears in the figure of an awful old woman.

Later, the soul must cross the Cinvat Bridge, which stretches through an unfathomable abyss. This bridge, which connects the Earth with heaven, widens under the feet of the righteous, but becomes narrow as a razor blade when an evil person tries to cross it. As a result, good people manage to reach the end of the trip, where they are received by Ahura Mazda and fed with milk and butter. On the contrary, the wicked fall into the abyss and go to the abode of Angra Mainyu, who makes them be served unpleasant and poisonous substances.

Centuries after Zoroaster, when his doctrine became an official religion, first of the Parthians, then of the Sassanid Empire, the original concepts had undergone important changes that accentuated dualism. Ahura Mazda, the principle of good, had become Hormuz, while his opponent, Angra Mainyu, the principle of evil, was renamed Arhiman and considered equal in strength and immortality to Hormuz, although at the end of time he'll be finally destroyed, along with his supporters, in Ahura Mazda's fire.

Judaism, which was born in the Middle East during the second millennium BC and reached the highest peaks of its development during the first, shared initially the Mesopotamian beliefs about the netherworld (the *Sheol*), as a place where the souls of the dead experience a sad and dark existence: *For them, love and hatred and rivalry have long since perished. Never again will they have part in anything that is done under the sun... Anything you can turn your hand to, do with what power you have; for there will be no work, no planning, no knowledge, no wisdom in Sheol where you are going*[185].

From the second century before Christ, a new conception of the future life appears in the biblical scriptures. Although it appears incipiently in very early writings, until that time it wasn't analyzed properly. In this new conception, after the present life there will be a judgment and a retribution: a reward for the righteous, a punishment for the wicked. There is also the idea of the resurrection of the dead at the end of the world, which in the time

---

[185] *Ecl.* 9,6-10.

of Christ had not yet been accepted by all currents of Judaism, as the Sadducees. In the second book of the Maccabees, which tells of the martyrdom of seven brothers and their mother for refusing to carry out practices contrary to their religion, the tortured brothers comfort each other with the hope of being resurrected in a better world: *You accursed fiend,* exclaims one of the brothers addressing King Antiochus Epiphanes, *you are depriving us of this present life, but the King of the universe will raise us up to live again forever, because we are dying for his laws*[186].

Future life was now imagined as a return to paradise, to the renewed Earth or (more rarely) to God's dwelling place. The righteous will enjoy immortal life and bodily pleasures: there will be exquisite meals, women won't be sterile and will have thousands of children. The wicked, on the other hand, will be thrown into *Gehenna* (the name of a garbage dump outside Jerusalem), where they will be tortured by inextinguishable fire. But there are those who argue that the wicked will be annihilated during the final judgment and therefore the penalties of hell wouldn't be eternal. For others, they would simply be temporary penalties and God's mercy, finally triumphing over his justice, would bring all to Himself.

Many of these beliefs were copied centuries later by Islam, which also combined Christian and Mazdeist elements without adding anything original from the eschatological point of view.

* * *

So far we have briefly reviewed the ideas of the different religions about the destiny of man after death, leaving Christianity aside on

---

[186] II *Mac.* 7,9.

purpose. All those beliefs about the other world can be classified into the following groups:

1. Those that consider the next life as a mere continuation of our present life.

2. Those that describe the kingdom of the dead as a horrible place, usually underground, where most of the deceased suffer a rudimentary existence.

3. Those who identify the other life with a state more or less similar to unconsciousness and loss of personality.

4. Those that include remunerative elements as a result of a trial possibly presided by a superior being, usually God, or a god. The believers in this group accept the existence of two kingdoms of the dead:

    • That of the righteous, which may or may not be the same as the abode of the supreme God, is imagined as a place of joy and delight. The joys of the blessed, if described, are usually simple extrapolations of earthly pleasures (exquisite food, beautiful women, etc.). This paradise is usually located, either on the Earth itself (on mysterious and remote islands) or in the future, either on the physical sky (imagined as a solid surface) or on other stars.

    • The second kingdom of the dead is hell, the abode of the wicked, imagined as a place of horror and torments. The sufferings of the damned, if described, are usually extrapolations of earthly pains (illness, despicable foods, physical pain, etc.). Hell is usually placed, either in the netherworld, or in other stars. There are hot and cold hells. Fire is one of the main instruments of torture in the former. Popular Christianity retains many elements of this.

Both paradise and hell are located in space and are subject to time. They are, therefore, a part of this universe. Divine eternity and the immortality of the human soul are imagined, in these religions, as an indefinite duration.

＊＊

Let us now turn to Christian eschatology. The first original concept of Christianity, the distinction between *eternal* and *everlasting* (understood as simple temporal continuity) did not arise as a consequence of religious speculation, but is linked to the Greek philosophical thought of Plato and Aristotle. In his work *On Heaven*, Aristotle says: *Outside heaven there is no place, no emptiness, no time. Therefore, what exists there does not take up space, nor is affected by time.* Aristotle places in that *spiritual center of the universe*, outside the astronomical sky, *the motionless motor* or *the God of Philosophy*[187].

The concept of eternity passed to Christian theology through Platonic and Aristotelian philosophies. In the sixth century after Christ, the Christian Anicius Manlius Severinus Boethius (c. 480-524), minister of the Ostrogoth king Theodoric, discussed these philosophical issues in his *magna opus, Consolation of philosophy*[188], written after falling in disgrace, a process that ended in his execution. For Boethius, perpetuity (everlasting time) is an

---

[187] *Metaphysics*, book XII, chapters 6-7.
[188] *De Consolatione Philosophiae*.

indefinite succession of moments, each of which is lost as soon as it has been reached. Eternity, on the other hand, is *the timeless fruition of an unlimited life*. God is eternal, not perpetual or everlasting. He does not foresee the future, He does not remember the past, He simply sees them.

The medieval model of the world incorporated these concepts, first through Boethius and Augustine, later directly from Aristotle. The great scholastic thinkers of the 13th century, Albertus Magnus (1206-1280) and Thomas Aquinas (c.1225-1274), combined the philosophical theories of Aristotle, recently re-discovered in Western Christianity, with the development of a new theology. In this way, Christianity was the first to assign to God the attribute of eternity, in the philosophical sense of the term. At first, however, it was considered that all created beings, including angels and the celestial spheres (following the Ptolemaic image of the world, then prevalent), lacked this divine attribute, although as immortal beings they were everlasting[189].

The second innovation introduced by Christianity in the context of the future life dates back almost to the origins, and reaches maximum development in the writings of St. Paul[190]. It is the doctrine of *the mystical body of Christ*: Christians, together with Christ, make a unique body in which Christ is the head and each of us members. This body, at present, is just partially formed: there is much still missing, before it can unite all the monads of the conscious universe that will freely agree to participate in it. When

---

[189] About the medieval model of the world, see *The Discarded Image*, 1964, by C. S. Lewis.
[190] I *Cor*, 12,12-27. See also *Rom*, 12,4-5. I *Cor*, 10,17. *Ef*, 4,4. *Ef*, 4,15-16.

this is done, creation will have reached fullness. Everything will be consummated in what St. Paul calls the *pleroma*, where the righteous will attain eternal happiness.

The force that brings together and keeps united the mystical body is love, which we can define as the consciousness of the members of their union in origin, organization and destiny with the remaining components of the mystical body. This love is the fundamental condition and energy[191] that enables the mutual help of all the members. The obligation to help others is an immediate consequence of being members of the same body. God himself participates as a distinguished member (the head) through Christ. For this reason, the two fundamental commandments of Christianity are mandates of love: *You shall love the Lord, your God, with all your heart, with all your soul, and with all your mind... You shall love your neighbor as yourself*[192].

The conjunction of these two innovations led to the application of the attribute of eternity to the mystical body (because God is a part of it) and to each of its members (by participation). According to this perspective, the members of the mystical body must one day abandon the space-time where we now live in the universe and move on to eternity. After the resurrection, at the end of time, being part of the mystical body will be equivalent to what has traditionally been called *being in heaven*.

The concept of hell, the destiny of the damned, has changed in the Christian mentality over time. From the literary and traditional vision of an underground hell (as in *The Divine Comedy*), we have

---

[191] *L'énergie Humaine*, Pierre Teilhard de Chardin.
[192] *Mt*, 22,37-39.

moved on to the modern concept, which identifies it with a state of the soul, rather than a place in space[193]. We now tend to think that salvation and condemnation, rather than the consequence of a trial, is a voluntary and conscious choice of each person: those are saved who renounce their own will, putting God at the center, as the head of the mystical body through Christ, while those who refuse to accept this and put themselves at the center, are in hell.

In the words of C.S. Lewis, in his book on this subject[194]: *In the long run the answer to all those who object to the doctrine of hell, is itself a question: "What are you asking God to do?" To wipe out their past sins and, at all costs, to give them a fresh start, smoothing every difficulty and offering every miraculous help? But He has done so, on Calvary. To forgive them? They will not be forgiven. To leave them alone? Alas, I am afraid that is what He does.*

In this version of the Christian world vision, after death we'll become a part of the mystical body of Christ. Heaven is the state of those who integrate into that body, accepting that God is the center and submitting their will to the Divine. Hell is the situation for those who don't want to accept it and insist on continuing to seek their own benefit against the common good of the whole. Purgatory would be the process by which certain individuals, starting from a strong opposite inclination, fight themselves until they accept integration. The righteous are equivalent to the cells of

---

[193] When Pope John Paul II said this in one of his speeches, the media, as usual, didn't understand. There was an immediate babel of voices that held that the Pope had denied the existence of hell. Although the Church discarded this interpretation, years later the idea the media planted in society continues to resurface from time to time.
[194] *The problem of pain*, 1940.

our body, which renounce an independent life and let themselves be controlled by the central nervous system (the head). The damned are like cancer cells that isolate themselves, encyst and don't want to have anything to do with others.

* * *

By following the evolution of the universe since its origin, through the four great stages through which life has passed on Earth until now, the extrapolation of the evolutionary processes that have acted in the past led us to predict the future emergence of the fifth level of life, of which incipient examples exist today. Logic and the study of human and social nature led us to foresee some of the main characteristics of the future fifth level being.

Surprisingly, we have found the same characteristics in an entity with a completely different origin: St. Paul's *pleroma*, whose oldest description dates back since the first century of the Christian era, almost two millennia before the scientific discovery of evolution. The cellular structure of the mystical body is quite clear. Saint Paul speaks of members rather than cells, because the latter were unknown in his time.

In our scientific study, we have observed that the body of the fifth level, in its current state, still has no head. The *pleroma*, however, does have it: God in Christ. But the union of the head with the rest of the body has not been consummated, it won't take place perfectly until the end of time. Therefore, the missing head is already ready, and is nothing less than God Himself.

The similarity of properties between the *pleroma* and the fifth level of life is evident: solidarity between its members, unity through love, dissociation from time... In addition, there will only be a single mystical body. Even in the problem of hell we have reached quite similar conclusions. Is it possible that all these concepts: St. Paul's mystical body, Teilhard de Chardin's *Omega point*, the fifth level being, are all equivalent? Is it possible to arrive to the same end of the trip by two totally independent paths, one scientific, the other religious?

In Chapter 9, we left unexplained the real reason why Teilhard de Chardin called the goal of evolution *the Omega point*. Now we can say it: according to the Bible, God is *Alpha* and *Omega*, the beginning and the end of the universe. At the end of evolution, the thinking universe will unite with God *Omega* in the person of Christ, the head of the mystical body.

The almost absolute coincidence of the *pleroma* with the fifth-level being moves us to identify them and lends enormous coherence to the image, both scientific and Christian, of the world around us. The universe is not meaningless, as nihilists insist. The eternal questions: *Where are we going? Where do we come from?* do have an answer. Science and religion do not disagree. Each of them has its own field of action and its own methods, but their scopes are not totally disjoint: they converge in their predictions about the future of evolution.

Manuel Alfonseca

Bibliography

1. Evolution of the universe before the apparition of life

Bondi, H., *Cosmology*, 1970.

Gonzalo, J. A., *Cosmic paradoxes*, World Scientific, New Jersey, 2012.

Guth, A. H., Steinhardt, P. J., "The inflationary universe", *Scientific American*, May 1984.

Hawking, S., *A brief history of time*, Bantam Books, 1988.

Rees, M., *Just six numbers*, Basic Books, New York, 2000.

Weinberg, S., *The First Three Minutes*, Bantam Books, 1979.

2. The first level

Alfonseca, M., *La vida en otros mundos*, Mc-Graw-Hill, Madrid, 1993.

Bada, J. L., *Cold start*, The Sciences, May/June 1995.

Diener, T. O., "Viroids", *Scientific American*, January 1981.

Ricardo, A., Szostak, J. W., "The origin of life on Earth", *Scientific American*, September 2009.

Smith, J. Maynard and Szathmáry, E., *The major transitions in evolution*, Oxford University Press, 1995.

3. The second level

Crick, F. H. C., "The genetic code III", *Scientific American*, April 1966.

4. The third level

De Duve, C., W. "The birth of complex cells", *Scientific American*, April 1996.

Margulis, L., Dolan, M. F., "Swimming against the current", *The Sciences*, January/February 1997.

5. The fourth level

Álvarez, W., Asaro, F., Courtillot, V. E., "What caused the mass extinctions?", *Scientific American*, October 1990.

Ayala, F. J., "The mechanisms of evolution", *Scientific American*, September 1978.

Bergson, H., *L'evolution Creatrice*, 1907.

Fortey, R., *Life: an unauthorised biography*, Harper Collins, 1997.

Gould, S.J., *Wonderful life*, 1989.

Mayr, E., "Evolution", *Scientific American*, September 1978.

Schlichting, C.D., Mousseau, T.A. (editors), *The year in evolutionary biology 2008*, Annals of the New York Academy of Sciences, vol. 1133, Boston, 2008.

6. What is man?

Cavalli-Sforza L. and Feldman M. *Cultural versus biological inheritance: phenotypic transmission from parents to children.* Human Genetics 25: 618-637, 1973.

Cloak, F. T., "Is a cultural ethology possible?" *Human Ecology* 3: 161-182, 1975.

Dawkins, R., *The selfish gene*, Oxford University Press, 1976.

Dobzhansky, T., *Human culture: a moment in evolution*, with Ernest Boesiger, Columbia University Press, New York, 1983.

Huxley, J., *Man stands alone*, Harper, New York, 1941.

Simpson, G.G., *This view of life*, Harcourt, Brace & World, 1964.

Turner, M. S., "More than meets the eye", *The Sciences*, November/December 2000.

Tudge, C., *The variety of life*, Oxford University Press, 2000.

Wikipedia: *Cladistics*, https://en.wikipedia.org/wiki/Cladistics.

7. Towards the fifth level

Alfonseca, M., *Human Cultures and Evolution*, Vantage Press, New York, 1979.

Childe, V. G., *Man makes himself*, The Rationalist Press Ass., 1936.

Gordon, D. M., "Close encounters", *The Sciences*, September/October 1999.

von Frisch, K., *The dancing bees*, 1927.

8. The fifth level in literature

Lewis, C. S., *The Abolition of Man*, Macmillan, New York, 1947.

Garaudy, Roger, *L'alternative*. English translation *The Alternative Future*, Penguin Books, Middlesex, England, 1976.

Kroeber, A. L., *Configurations of Culture Growth*, University of California Press, 1969.

Sorokin, P. A., *Social and Cultural Dynamics*, Porter Sargent, Boston, 1970.

Sorokin, P. A., *Society, Culture and Personality*, 1947.

Spengler, Oswald, *Der Untergang des Abendlandes*, 1918-22. English translation *The decline of the* West.

Toynbee, Arnold J., *A Study of History*, 1934-1961.

Weber, Max, *Die protestantische Ethik und der Geist des Kapitalismus*, 1905. English translation *The Protestant Ethic and the Spirit of Capitalism*, 1930.

9. The Omega point

Tobias, P. V., "Piltdown unmasked", *The Sciences*, January/February 1994.

10. Internet as a nervous system

Penrose, R., *The emperor's new mind*, Oxford University Press, 1989.

Sáez Vacas, F., *Más allá de Internet: la red universal digital*, Centro de Estudios Ramón Areces, 2004.

Sagan, C., *The Dragons of Eden*, Random House, New York, 1977.

11. Must we renounce reproduction?

Alfonseca, M., de Lara, J.: "Two level evolution of foraging agent communities", *BioSystems*, Vol. 66:1-2, p. 21-30, June/July 2002.

Bak, P., *How Nature works*, Oxford University Press, Oxford, 1997.

Farrell, W., *How hits happen*, Harper Collins, London, 1993.

Heylighen, F., Campbell, D. T., *Selection of organization at the social level: obstacles and facilitators of metasystem transitions*, publicado en *World futures: the Journal of General Evolution*, vol. 45, p. 181-212, 1995.

Mantegna, R. N., Stanley, H. E., *An introduction to Econophysics*, Cambridge University Press, 2000.

Waldrop, M. M., *Complexity*, Penguin, 1992.

Ward, M., *Beyond chaos*, Thomas Dunne Books, 2002.

12. Can we control our evolution?

Birney, E., "Hidden treasures in junk DNA", interview by Stephen S. Hall, *Scientific American*, October 2012.

Mullis, K. B., "The unusual origin of the polymerase chain reaction", *Scientific American*, April 1990.

Watson, J. D., *DNA: the secret of life*, Alfred A. Knopf, 2003.

13. Should we control our evolution?

Wuethrich, B., "All rights reserved", *Science News*, 4 September 1993, pg. 154-157.

14. How will the fifth level be?

Webb, S., *Where is everybody?*, Praxix Publishing, 2002.

15. Does the fifth level exist?

(Anonymous), *Egyptian book of the dead*.

Alfonseca, M., *Krishna frente a Cristo*, Madrid, 1978.

Eliade, M., *A history of religious ideas*, Collins, 1979.

James, E. O., *Teach yourself history of religions*, The English Universities Press, 1956.

König, F. et al., *Christus und die Religionen der Erde*, Verlag Herder, Wien, 1956.

Puech, H. C. et al., *Histoire des religions*, Gallimard, 1970-72.

Smith, H., *The world religions*, Harper San Francisco, 1991.

www.ingramcontent.com/pod-product-compliance
Lightning Source LLC
Chambersburg PA
CBHW071350210526
45465CB00001B/43